FATEFUL
HARVEST

FATEFUL HARVEST

*The True Story of a Small Town,
a Global Industry, and a Toxic Secret*

Duff Wilson

HarperCollins*Publishers*

HarperCollins books may be purchased for educational, business, or sales promotional use. For information, please write: Special Markets Department, HarperCollins Publishers Inc., 10 East 53rd Street, New York, NY 10022.

Maps by James McFarlane

FIRST EDITION

Book design by Lisa Sloane

Printed on acid-free paper

Library of Congress Cataloging-in-Publication Data
Wilson, Duff.
  Fateful harvest / Duff Wilson.—1st ed.
    p. cm.
  Includes bibliographical references.
  ISBN 0-06-019369-7
    1. Fertilizer industry—Environmental aspects—Washington (State)—Quincy. 2. Factory and trade waste as fertilizer—Environmental aspects—Washington (State)—Quincy.
3. Hazardous wastes—Environmental aspects—Washington (State)—Quincy. 4. Fertilizers—Health aspects—Washington (State)—Quincy. I. Title.
  TD195.F46 W55 2001
  363.19'2—dc21   2001024208

01 02 03 04 05 CG / RRD 10 9 8 7 6 5 4 3 2 1

For my wife, Nancy,
And children, Lana and Grant

A NUMBER OF VOICES (SHOUTING):
Don't talk about the Baths! We won't hear you! None of that!

DR. STOCKMANN:
Do you imagine that you can silence me and stifle the truth! You will not find it so easy as you suppose.

It is I who have the real good of the town at heart! I want to lay bare the defects that sooner or later must come to the light of day. I will show whether I love my native town.

HENRIK IBSEN
An Enemy of the People (1882)

Topsoil is more vital to human survival than almost any other resource, for without topsoil we cannot feed ourselves.

MICHAEL S. NORTHCOTT

Th' newspaper . . . comforts th' afflicted, afflicts th' comfortable, buries th' dead an' roasts him afterward.

FINLEY PETER DUNNE

The events in this book are true. Every word inside quotation marks was heard by the author firsthand or on tape; reconstructed dialogue is shown in italics.

CONTENTS

Sites from Fateful Harvest

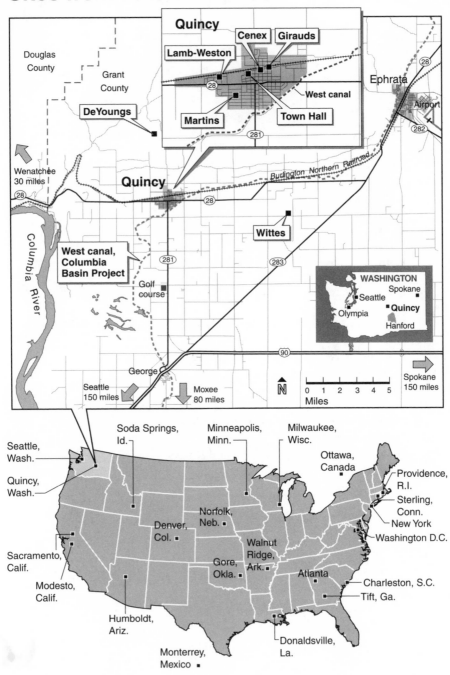

BY DUFF WILSON
Seattle Times staff reporter

When you're mayor of a town the size of Quincy, Wash., you hear just about everything.

So it was only natural that Patty Martin would catch some farmers in her Central Washington hamlet wondering aloud why their wheat yields were lousy, their corn crops thin, their cows sickly.

Some blamed the weather. Some blamed themselves. But only after Mayor Martin led them in weeks of investigation did they identify a possible new culprit: fertilizer.

They don't have proof that the stuff they put on their land to feed it actually was killing it. But they discovered something they found shocking and that they think other American farmers and consumers ought to know:

Manufacturing industries are disposing of hazardous wastes by turning them into fertilizer to spread around farms. And they're doing it legally.

"It's really unbelievable what's happening, but

FATEFUL
HARVEST

Prologue

THE ROAD TO QUINCY crosses the Columbia River and rises to the greening fields where fortunes are won and lost on the whims of markets and the weather. Farms line the highway. Farm dogs lie on driveways soaking up the sun. The fields are marked with roadside signs: BEANS . . . CORN . . . POTATOES . . . WHEAT. The people here like to say they feed the world.

My car speeds toward the small town in the distance. State Route 281 goes straight as a survey line. I drive in the quiet, windows open, sun high. The air smells fresh with life soon to bloom. Only an occasional pickup truck shares the road with me, the newspaper reporter slipping into town to watch the mayor of Quincy in action in enemy territory: the town that elected her.

I'm going to see them try to shut her up.

I remember my surprise at first meeting the small-town mayor. I imagined someone like my wizened great-aunt. I hadn't expected Patty

Martin to be young, tall, and an attractive mother of four. And I hadn't expected her to make so many accusations against so many pillars of her community.

She claims toxic chemicals are being turned into fertilizers and slowly poisoning some of our food. In her telling, many of the worst chemicals known to twentieth-century industrial pollution—arsenic, beryllium, cadmium, dioxins, and so on through the alphabet of deadly waste—are sneaked into common farm and garden fertilizers. Hazardous waste is mixed with plant food, she says, to save industry money instead of being buried, burned, or purified safely.

And nobody knows about it. That's the kicker. Nobody knows.

It was only natural this claim would threaten the people in Quincy. Martin says the toxic chemicals are plowed into farm fields and tilled in gardens by the millions of pounds, ending up in plants and animals. In her town, and everywhere, the mayor says, companies put almost anything in fertilizer if there's a little nutrient like phosphorus or zinc along with the unknown ingredients in the mix. She says it is recycling run amok. It's the ultimate dilution solution: What they can't dump in the oceans or pour in the air, they mix with dirt. Fertilizer is an easy place to hide toxics behind the "green mask of recycling," she says, because the protectors of public health and product safety don't look any deeper than the advertised ingredients.

She has a voice like a coach or sergeant, heavy and booming, and it matches her unexpected height and outsized head and hands.

"It's really unbelievable what's happening, but it's true," she says. "They just call a dangerous waste a product, and it's no longer a dangerous waste. It's a fertilizer."

Much of this was unbelievable, but I liked her attitude. She was bold. Martin had told me she was a threat to people with fortunes to be made recycling hazardous wastes, as well as a threat to a fertilizer industry with sixty-billion-dollar sales around the world. The chemical companies that ran Quincy were out to destroy her. I could understand why.

She'd said she needed my help. While intrigued, I was also skeptical. With twenty years' newspaper experience, I know how people can misuse anecdotes and statistics to draw false conclusions. Martin seemed too quick to see a cause where none was proved. But the scheme, as she'd explained it, also made a perverse sort of sense. Who in the world would think to look for toxic waste in plant food?

I see the soil in a new light, and I wonder about my own lawn and garden. What have I sprinkled in my backyard? Is somebody using my home, my food, to recycle toxic waste? It seems unbelievable, outlandish—but what if it's true?

The woman I'm driving to see in action doesn't fit the profile of an environmental extremist or conspiracy theorist. Martin is a conservative middle-class housewife, the daughter of an Air Force colonel and a nurse. She once played professional basketball in Europe, then married a man who manages frozen-food production. She'd started a recycling program in her hometown before being pushed into running for mayor, a part-time job paying five hundred dollars a month.

She stumbled on a secret and started asking too many questions. Affection for her in the town has turned to hostility and fear: She is considered dangerous. Martin says the people who made money deceiving farmers and gardeners would do almost anything to keep their secret. She suspects her phone is tapped and that she's spied on. She thinks her life is in danger.

WELCOME TO QUINCY

Twenty-five years before, I'd left a one-stoplight town with four thousand people like the one I am approaching today. I feel like I'm going home. I grew up a hundred miles to the north, in Omak, Washington, with ink in my veins: My parents edited and published the weekly *Omak Chronicle*.

I drive past signs advertising the Quincy Rotary Club and the Lions Club. It's true what they say about small towns, the closeness, the community. It comes out in good ways when somebody needs help or somebody dies and people gather round. They don't look away. But it can come out in bad ways, too. I think about the courage it takes to challenge a power structure like the fertilizer industry in a farm town.

My own instinct had been to flee. I'd moved to the coast and become an investigative reporter at the *Seattle Times*. Investigative reporters don't wait for news to happen; they dig it up. We like to think we help people, but a lot of them don't seem to think so. You can't do a lot of investigative reporting in a small town. You'd upset too many businesspeople. I wondered how Mayor Martin could get away with troublemaking in her town.

As I approach Quincy, I see a reader board in the pale blue sky over a building marked with the red-and-white sign of Cenex. "Council Meeting Today, 2:30." It is clear who runs this town.

Cenex, a subsidiary of Cenex/Land O'Lakes Company of Minneapolis, sells four billion dollars a year in seed, fertilizer, fuel, animal feed, pesticides, credit, and advice to farmers in most of the nation. The railroad tracks in Quincy are lined with a string of chemical and food-processing companies. Cenex/Land O'Lakes is the knot at the center of the string. Cenex is the farmer's best friend.

Yet it was from Cenex in Quincy that Martin first picked up the idea of calling hazardous waste fertilizer.

The company had stored its waste from field operations, laced with heavy metals, in a concrete pond by the railroad tracks. Cenex dug a bigger pond than most other companies that had similar storage. Cenex evaporated the liquid by pumping it up into the sky like a fountain. The spray turned to mist that children walked through on their way to school.

What was left in the pond was thirty-eight thousand gallons of toxic waste. Martin says Cenex/Land O'Lakes spread the waste on a farmer's land by calling it a fertilizer "product" instead of putting it in a hazardous-waste dump, where it belonged. She claims it contaminated the land, poisoned hay, killed a neighbor's horses, and that Cenex lied about it. Then the mayor says Cenex took over the farm and watered it hard to flush the toxics out of the soil. This is where she got the idea something was wrong.

Call me naive—I thought Land O'Lakes just made butter—and I was surprised by the story she'd told me. More than surprised. I couldn't believe it.

Martin also claims a subsidiary of Alcoa, the largest aluminum company in the world, sells a dangerous industrial waste as fertilizer or road deicer. Whatever you needed: Ag-Mag to fertilize fields, or Road Clear to melt ice. Different labels, same material. Sold all over the northwestern United States and western Canada. Martin says the company would have to pay two million dollars a year to put the material in a dangerous-waste landfill if it didn't recycle it as a "product."

"It's nothing more than hazardous waste that melts, then dries, then blows in the wind," she'd said, and provided evidence on company letterhead for each example.

I park my car on a dusty street and walk into the town hall and a rancorous scene. Three tables are pushed together for the mayor, council, and city attorney at the front of a small meeting room. I immediately see Mayor Martin, the only woman at the front of the room. Sixty other people fill eight rows of folding chairs at the back and dozens more stand in the hallway. It's the most crowded meeting the council had held in years, and in the middle of a workday afternoon. The Cenex managers have told their employees to go, and they are dirty with the day's work behind them, and angry at the mayor.

A lot of them have friends or family who lost money in food scares, especially after the *60 Minutes* treatment of the Alar story. Some of them say their jobs are on the line today. The mayor is talking about secret toxics in the chemicals they use to grow food. She is talking about lead in french fries they sell to McDonald's. They take it personally.

Somebody hands out a letter warning: "A group of farm operators and food processors are prepared to file a lawsuit against the Mayor and the City if these unfounded allegations continue."

At the front of the room, Martin stands to speak. The townspeople rustle to their seats. At six feet and one-half inch in height, Martin stands taller than most of the men in the room. At forty-one, she looks ten years younger, slim from swimming and other athletic pursuits. Her brown hair flows past her shoulders like a schoolgirl's. She looks uncomfortable in her dress, like she'd rather be wearing jeans. And like she'd rather be anywhere but here.

"I want everybody to know," she says, "there has been no attempt on my part to do anything to harm the farmers in this community."

Somebody snorts. Other people stare coldly.

Martin is trying to help some of the poorest farmers in the area, but they are neither standing nor speaking in her defense today. Far outnumbered, they don't want to pour diesel on the fire.

Dennis DeYoung is her unlikely best friend. Dennis had once been one of the most successful farmers in the county, worth more than five hundred thousand dollars, but lost everything. Now he mills grain, drives a truck, and lives in a mobile home on his parents' land. Dennis has sad, watery blue eyes and a cackling laugh. He is quiet today. He has four children and a most unhappy wife.

Duke Giraud, onion grower, is days from losing his food-packing business. Duke and his wife, Jaycie, are deep into bankruptcy and mad as hell about it. Martin has everything to lose from helping the Girauds.

And Tom Witte is a dairy farmer with cancerous cows. Tom is so poor he can't afford more than fifty gallons of tractor gas. Like the others, he is trying to pin the blame on contaminated fertilizer.

Martin tells the townspeople why she is stirring this up. Her concerns are many and serious. She is worried about lung disease, cancer, and birth defects. She is afraid they are all being put at risk by toxic metals added to the soil and blowing in the dust. Not just in Quincy. Everywhere.

"I don't consider myself a champion of a cause," she says. "I consider myself a person who happened to stumble on that information, and it was bigger than I could handle and so I passed it along. It was just my intent to make the farmer aware of the product he was getting."

The grower who'd written the lawsuit letter rises from his folding chair. He wears a spotless white cap that sets off his face, flushed red and angry. "What we've seen coming out is the start of another Alar scare," he says. "We had a woman starting that one, too."

Many in the audience laugh.

An orchard owner holds up a flyer from the state department of health. Almost shaking with rage, she turns to Martin. "There's no reason to scare people. It's going to create hysteria!"

A council member, Tony Gonzales, turns to another out-of-town guest standing at the side of the room: a man from the state health department. "Why did the state get involved?" Gonzales says. "You're down here talking to citizens about their health problems, but you don't live here. You're talking about people's livelihood here. You're talking about people who make a living growing crops."

Gonzales says it's time for the mayor to stop asking reckless questions and start following council orders.

Martin knows him well. Tony Gonzales works with her husband, Glenn Martin, at the food plant. Tony used to encourage Patty to confide her suspicions in him, then he turned against her. He is the son-in-law of the local manager of Cenex.

The president of the Washington State Potato Commission, also a Quincy resident, rises to speak. He has the cocky manner of a man who's used to having his coffee cup refilled frequently and his advice followed. He wears sunglasses perched atop his thick, wavy hair, and speaks with confidence.

"I'm sure there's a quieter way to deal with this than to go to the press with it," he says. "You *gentlemen* understand. *We* have a problem within our *own* nest that we need to deal with."

As if on signal, the city attorney tells the council to close the rest of the meeting to the public. He holds up the letter threatening a lawsuit. "I would recommend the city—the five gentlemen and the mayor—discuss this in executive session before we start airing out things," the lawyer says.

So the public meeting ends. The council plans to call a town hearing the next evening, but now they'll talk privately with the mayor. The citizens file outside, most going home to dinner. Only five councilmen, the city attorney, and Mayor Patty Martin remain in the room. They close the door to the hallway. They close the front door to the building.

I wait outside. It's a beautiful spring afternoon with songbirds lilting in the azure sky and wind rustling the trees.

Before long I can hear the men through two closed doors. They are finally getting down and dirty. They are yelling at Martin to stop— before it's too late.

"I told you!" one of the men shouts, and pounds the table. "Be quiet!"

CHAPTER I

SMALL-TOWN STORIES

ONE NIGHT A DECADE EARLIER, as farm families were settling down in homes set back from the highway, Patty Martin drove across the bridge spanning the Columbia River and up to the plateau leading to the Quincy Valley. Cows stood quiet in the gloaming. The Milky Way glimmered in the sky. Patty could see a row of lights marking Quincy from five miles up the road.

Patty was coming home, after years far away, home again to Quincy, bringing a husband and two healthy, brown-eyed children. The cat was going crazy in Patty's car, but the children, five and two, were asleep. Glenn Martin followed in a truck with Shep, the family dog, lying quietly on the seat beside him.

Patty had spent much of her life on the move. She'd been born on Hamilton Air Force Base in California on November 6, 1956. Alfred and Erika Naigle had four sons and then Patty, followed by two other daughters. Al Naigle was a Strategic Air Command radar squadron commander. The family had moved every two years or so to bases in California, Mississippi, British Columbia, and Washington State.

Erika Naigle, a registered nurse, had done most of the child rearing while the commander came and went from assignments overseas. While Patty had balked, mulelike, at her mother's attempts to discipline her, she simply adored her father. He was a perfectionist. He had a place for everything: military honors sorted into cases, shoes lined up in the closet, stamps mounted in an immaculate collection. After Patty's mother volunteered as a Camp Fire leader for the girls and Cub Scout den mother for the boys, her father took over as a Boy Scout leader and district commissioner. Two of the four boys became Eagle Scouts.

When Al Naigle retired from the Air Force with lieutenant colonel's wings in 1963, he eventually found a sun-bleached, safe little town in which to settle down with the family. Quincy was a nineteenth-century rail stop. It had served dryland wheat farmers and well-water orchardists scratching a living out of the dust until Roosevelt and Truman tamed the galloping river that had amazed Lewis and Clark; then Quincy became a far more prosperous town.

The U.S. Bureau of Reclamation corralled the Columbia into narrow lakes bounded by basalt cliffs and concrete walls, most famously the Grand Coulee Dam, the largest man-made structure in history into the 1940s.[1] The water crashed through turbines to spin the cheapest electricity in the world, helping America win World War II. The power was wired to aluminum smelters to make airplane skin for Boeing B-17s and to Hanford Atomic Works to make plutonium for the A-bomb. After the war, the bureau put the Columbia to peaceful use, pumping water behind the Grand Coulee to flood a remote valley, twenty-six miles long and one mile wide. Banks Lake, it was called. Two hundred and fifty-six feet higher in elevation than the river, it installed gravity power for irrigation flow in the greatest farming project to that time. The bureau's plan for the Columbia Basin Project called for two great canals to flow out of Banks Lake, but the money ran out after one was finished.

The West Canal ran through the Quincy Valley.

Here the water was channeled to ever-smaller canals and pipes to quench the desert soil. Irrigation brought life to a deserted area the size of New Hampshire. The settlers enjoyed the cheapest water in the nation. The farming was intense with high rates of fertilizer, pesticides, and fumigants. An average American farmer feeds fourteen people and manages by far the most productive enterprise in the eleven-thousand-year history of farming. Columbia Basin farmers grew potatoes, alfalfa, corn, wheat, apples, seed crops, asparagus, and grapes for wine. Near the highway, the water ran pure and cold in unlined ditches.

This was where the Naigles called home. The Stars and Stripes flew from porches every Fourth of July. The community celebrated a farmers' day after harvest. At Christmas the fields and lawns were blanketed with snow, and gifts covered the floor under decked-out Douglas firs. Al went to work for the Bureau of Reclamation.

The Naigles had one goal in mind after years of packing and unpacking. They wanted a family place. They wanted all seven of their children to graduate from the same high school.

Patty always felt like the youngest boy, not the oldest girl, in the Naigle family. She had attended four schools by the fourth grade, invariably the youngest and tallest girl in class. By high school, Patty stood a head up on the other girls. She always felt different. She didn't have close girlfriends. She usually hung out with the boys. It was John and Fred and Ron and George and Patty. When the school stopped offering girls' track, Patty called the Superintendent and asked about Title IX entitlements for women's sports. He told her to turn out with the boys's track team. There she set the school record in the javelin. She was also president of the science club and vice president of the honor society. Patty had a steady boyfriend who was also a good student and athlete, of course. John Omlin and Patty Naigle were king and queen of the Sweetheart Dance.

"I'm the luckiest guy in the whole school," John wrote in her yearbook. "Who else do you know that has a girlfriend that is really good looking, intelligent (except when it comes to choosing a boyfriend), has a good sense of humor, has a nice personality, and who likes to turn out for sports. And also likes to get involved in activities instead of letting them go down the drain. Even though at times you can be as stubborn as a mule."

Patty was a standout basketball player from as early as she can remember. She was known for a smooth right-handed hook shot. The coaches built plays around her shooting skills. She was big and strong and unflappable when the whistles blew against her. She would shoot. Miss. Shoot again. Hit. Fifty percent was a good shooting percentage for basketball, and Patty Naigle also rebounded relentlessly.

But when she went to Gonzaga University in Spokane, there was little money for women's sports scholarships. She set out to study biology with ambitions for medical school.

One day Patty was spotted shooting in the gym. The college coach asked her to join the team. Patty averaged twenty-seven points a game as a junior at Gonzaga. She was ranked the number fourteen women's player in the nation by the Kodak All-America judges. In her senior year, though, the team got a new coach who didn't like Patty as much. The coach told Patty to shoot her pet hook shot less. She did. She hated playing ball her senior year. Still, the graceful center left Gonzaga as the school's all-time scoring leader and with a degree in biology.

She got two job offers, one to coach and teach at a high school in Spokane, the other to play basketball in Borlänge, Sweden. Patty became the only American woman in the Swedish leagues. She got a free apartment and fifty kroners, about eleven dollars, a day. She didn't get the car they'd promised, and she didn't play as well as she'd hoped in the European version of the game, which didn't allow inside players like her to stay as close to the basket as she was used to doing before a three-second violation. Patty turned down an offer to come back for a second year.

Returning from Sweden, Patty wanted to go back to school for dental hygienist training. Her ambition to go to medical school had somehow been lost, and her parents told her they were done helping pay for her education. Over the summer, living at home, she managed the public swimming pool in Quincy. The lifeguards called her "Miss Efficiency," noting both her ability to get things done and her insistence they be done her way.

Patty had no idea where she'd work or what she'd do when summer ended. She ended up applying for an entry-level management position at Lamb Weston, a frozen-food company whose highest-profile products were French fries, onion rings and frozen fruit pies for McDonald's and Burger King.

That was the day she met Glenn Martin. Patty showed up for the job interview wearing a blue floral chiffon dress and with her hair carefully curled, perhaps more suitably attired for a morning at church than a day at the factory. She was introduced to Glenn as a former pro basketball player in Sweden.

Oh, did you partake of the Swedish delights? Glenn Martin teased.

No, and if that's what it takes to get this job—. Patty reached for the door handle. Patty thought Glenn was asking if she was as sexually adventurous as Swedish women were fabled to be; she disapproved, both of those kinds of adventures and that sort of question.

Just kidding! Glenn said. He was tall, dark haired, with green eyes, and, offended or not, Patty liked his smile right away.

The managers took the overdressed applicant into the onion-processing room, where her eyes burned and watered, affording them further hilarity. Patty got the job, and pretty soon she had Glenn Martin over for dinner.

Later he told Patty that he knew if he dated her, he'd marry her. Glenn was a food sciences graduate of the University of Washington department of fisheries. At thirty, he was seven years older than she and focused on his career. In his spare time, he flipped through magazines like *Food Processing* and *Chemical Engineering*.

Patty Naigle was unusually straitlaced and naive for a twenty-three-year-old woman. She didn't drink before she was twenty-one, didn't approve of movies that had sex or bad language, certainly didn't smoke or swear. She said "Crap!" when she was annoyed and "Holy cow!" when she was surprised.

Glenn was hooked. On a night out with other employees from the plant, Glenn and Patty talked all about politics and religion and found out they had nothing in common. Patty was a hawk on Vietnam, antiabortion, and Catholic. Glenn was a dove, pro-choice, and agnostic. But they got along. Patty hadn't dated much since high school. Her father had told her to never date anybody she wouldn't want to marry. After three months, Glenn proposed.

Holy cow!

Patty called home. *Mom, what should I do?*

Well, you can always say "Yes" and make it a long engagement.

Eight months later, Patty and Glenn married. Glenn stayed in Oregon; Patty moved to Seattle to attend the university. They were together on weekends.

The best offer to Glenn came from Pacific Pearl Seafoods in Dutch Harbor, Alaska, 1,950 miles from Seattle. It was owned by the same conglomerate that owned Lamb Weston, Dole Pineapple, Monterey Cheese, and other food companies. Glenn flew up to interview, flew back with a huge box of king crab, and sold Patty on the taste of Alaska. They lived three years in Dutch Harbor on the Aleutian Islands, one of the most consistently cold, windy, and wet places on earth.

Patty would remember them as her happiest years.

The sun shone past midnight in the summer, and they bought a boat to crab and shrimp and fish together on the Bering Sea. They took otherworldly picnics on remote rocky beaches.

Patty worked as a clerk at the Iluliak Family Health Clinic. Soon she was the pharmacy manager. After some quick training, she became the island's emergency dentist, willing and able to pull people's teeth when necessary. There was no doctor and no dentist on the island.

Patty's first baby was born in August 1982 and named David, after Michelangelo's statue she'd seen in Florence after playing ball in Europe. The baby was just starting to walk the next year when Glenn said he wanted to go back to the Lower Forty-eight. The holding company Amfac was getting out of the seafood business, and Glenn wanted to stay with the same family of companies by working for Lamb Weston in Oregon again. Glenn took a pay cut and Patty gave up her job when they left Alaska, but Patty had banked two years of her salary.

They bought a house on four acres outside of Hermiston, Oregon, in sagebrush land some people called the "triangle of death"—bounded by Umatilla Army Depot, the Arlington, Oregon, dangerous-waste landfill, and the Hanford Nuclear Reservation. Patty knew little of that. She taught swimming at the pool. Hermiston was three hours from Quincy.

During a family reunion on Father's Day 1985, Al Naigle grabbed his abdomen in pain. He hoped it was a hernia, but it was worse than that. He had kidney cancer. Seventeen years earlier, doctors had removed one of Al's kidneys, and he had told his oldest daughter that cancers do come back and that he wouldn't live to see his sixty-fifth birthday. He wasn't off by much. He was sixty-five when the cancer returned.

Patty desperately looked for cures. She wanted her father to try alternative treatments, garlic, vitamin supplements. She believed nutrition was the key to illness. But her father's oncologist dismissed the idea as quackery. Patty never forgave the doctor for that. Patty called 1-800-4CANCER for advice from the National Cancer Institute. She ran up $150 phone bills calling medical schools and chasing leads. Patty and her sisters and mother took Al Naigle to the University of Washington to get an experimental treatment, but he was too sick to be enrolled. The cancer had spread to his bones.

Al Naigle died just before his sixty-sixth birthday. Looking at her father in his open casket, Patty thought, *Move over and make room for me.* She felt his absence deeply. She felt guilty she'd been away. She didn't stop crying for a year. No one ever measured up to her father.

Glenn and Patty Martin moved back home to Quincy in 1987, two years after the death. Patty's mother and two sisters still lived in town.

Glenn took a new job with Lamb Weston. Patty raised babies, watched what her family ate, and read every article on cancer and health.

Patty taught swimming for the American Red Cross. Before long the town council was offering to pay the energetic housewife three hundred dollars a month for a part-time job as head of Quincy Community Activities Recreation and Education, or Q-CARE. She started a recycling program, an after-school latchkey program and adult literacy council.

Patty started volunteering in her son David's first-grade classroom, then tutoring a little boy in reading one day a week, and soon recruiting twenty volunteers to tutor thirty-two children, nearly a quarter of the Quincy first grade. In 1990, she was given the Outstanding Service Award from the Quincy Valley Chamber of Commerce.

It was about then that the trouble started, on an evening Glenn watched the kids and Patty went to the town council meeting to talk about a volunteer project.

Some consultants from an engineering firm were droning on about a plan to expand the industrial wastewater treatment plant, and Patty was halfway listening and doodling notes while she waited for her spot on the agenda. Patty started paying more attention when she realized they were talking about two new vegetable-processing plants joining the system in addition to her husband's company, Lamb Weston, which was then the only user. Then she heard the council start to talk about raising home owner water rates to subsidize the expansion of the industrial treatment plant.

Patty wrote a letter to the Quincy weekly newspaper asking people to come to the next council meeting and raise their voices against the rate hike. More than fifty people showed up. The council nervously postponed action. For the first time, Patty felt a little like Ralph Nader.

But the council had already sold the bonds for the expansion project. They couldn't undo it. Patty wrote the state attorney general to complain. She sat in on more council meetings.

Then the council began to talk about buying wasteland for the project. After the food industry waste filtered solids from its waste—often used as cattle feed—the liquid still had to be disposed of somewhere. They were looking at buying 1,595 acres by the golf course to pour it on.

That really caught Patty's eye. It was five miles from town. It had houses and barns they'd have to clear. It was already irrigated land, and they wanted to spray millions of gallons of wastewater out there?

Quincy's industries were going to produce a volume of wastewater you would expect from a city of three hundred thousand, but still, the 1,595 acres was far more land than they needed. And they were talking about paying $3.2 million, far more than the cost of most land in those days. Her husband and her husband's boss encouraged Patty to be a citizen watchdog.

Why were they so determined to put the wastewater on that particular land? Why not pick some dry land closer to town? Why buy so much? Why pay so much? Why?

None of this is making sense.

Then somebody told her about Dennis DeYoung.

Dennis was a third-generation farmer who had made a fortune in his twenties growing wheat while Patty was playing schoolgirl basketball.

Dennis's grandfather had farmed on the Minnesota side of the Red River Valley. His father had moved to the promised land of cheap water in Quincy in 1961, bought two hundred acres, then doubled it. Dennis had pulled on his first pair of rubber work boots before he was eleven years old and grew up helping to change the water lines, pick up rocks, feed the calves, and harvest crops of corn, beans, wheat, and alfalfa. He liked the work but also knew its hardship. When he was fifteen, he rushed out in the field where his only brother, Dale, couldn't get up after falling from a hay truck. Dale, seventeen, was left paralyzed.

Dennis had no trouble handling more of the chores and could tune an engine and turn a wire into a widget, but he did not excel in school.

Later, he discovered that he had a reading disability that could have been helped. He never got over the feeling, though, that he was a little bit different. After high school, Dennis worked on the farm and drove a truck for wages. Within three years, he saved enough money from his ventures to buy 480 acres of wheat land and lease 2,560 more in a nonirrigated area he called Sagebrush Flats.

The first year on his own, Dennis struck gold. Wheat prices shot up in 1973 when the United States started selling to grain-starved Russia. Dennis could have paid off his entire seventy-thousand-dollar mortgage. Wheat is a hardy crop, and Dennis and his neighbors got just enough rain, ten to twenty inches a year, most of it in the winter, for dry farming to produce forty or fifty bushels an acre. He had five more good years in Sagebrush Flats, bought a Cessna 140 to fly between his place and his parents' place and all over the West. But when his mother was diagnosed with ovarian cancer and given six months to live, Dennis sold his land to move back home. His father, Jake, was fifty-nine and needed help.

In June 1979, Dennis put one-third cash down on the $156,000 price of 140 acres out on Road 11 NW and Road U NW, four miles from Quincy and ten miles from his father's farm. Dennis's land overlooked a little lake and, farther north, a prehistoric drainage of channeled scabland called the Crater Coulee. Dennis's farm was one of the better pieces of ground that could be found for sale at that time.

Dennis's soil was rich, dark, deep, and well drained. It was better than clay, which doesn't breathe and fights a plow. It was better than sand, which doesn't hold liquids at all. Dennis's soil was loam, a mixture of textures, and as good as it gets. It was silty loam settled from the sea that had once covered the West. And it was deep. Nature takes five hundred to one thousand years to make an inch of topsoil from weathered rock and the decay of plants and animals. Dennis's soil was ten to ninety feet deep in most places. The surface of the land had a gentle roll, good road access, and most important, water rights.

On the Sagebrush Flats, Dennis had had to rely on the weather God sent him. In the Quincy Valley, he relied on the beneficence of Franklin

Delano Roosevelt. And what irrigation Roosevelt begot, the U.S. senators from Washington propagated in pork barrel projects. Dennis held water rights to enough water to cover his land five feet deep each year.

He used a single sprinkler pipe almost a quarter mile long. The length of the pipe was elevated fourteen feet on eight metal struts, each supporting a sprinkler tower above and mobilized by a set of rubber wheels on the ground. The Sargent Raincat system revolved around a center pivot, like the minute hand on a clock, powered by an electric motor. Each joint in the pipe contained a microchip and sensor to turn the wheels at different speeds, keeping the whole line straight.

Dennis was skilled and proud of his circle. He maintained all the equipment himself. When he flew overhead in the growing season, his field looked like a perfect green O. Flying higher, it was a dot among hundreds.

Every day, Dennis drove back and forth, back and forth, between his farm and his father's farm, working them both. He improved his land by burying twelve-inch plastic pipe in an open drainage ditch that ran through the middle and covering the ditch with dirt, and he planted red Delicious apple trees in eighteen acres of corner land. Dennis watered the apple orchard by hand line. He left fallow twenty-nine acres sloping to the lake.

Things were going so well, Dennis cosigned a forty-five-thousand-dollar partnership loan with his father to build a grain mill on the old family farm and assure a steady retirement income for his parents. He could buy grain from other farmers and run it through the mill to make animal feed and other products. It was the only grain mill in the local area.

And he'd begun to believe his mother, enrolled in an experimental cancer treatment program, just might live.

A year after he came back to town, Dennis married a pretty young woman named Marilyn Fenn. Eight years older than she, he had known her since she was sixteen. They had dated off and on, and now Dennis was twenty-nine and wanted to settle down. Marilyn, for her part, thought Dennis looked like the film actor Kurt Russell.

"It's his eyes," she said.

Marilyn was a farmer's daughter, fair and blond, and she wanted to be a farmer's wife. Dennis took her farther than she'd ever been before on their honeymoon, a month-long road trip with fourteen hundred dollars in cash. Marilyn sat over by Dennis, her left leg touching his right leg, while Dennis steered their new Ford pickup—it was beer bottle brown, they laughed—down the coast highway to California and over to deep-sea fish off Texas.

By now Dennis owned a jet-drive ski boat, a Harley-Davidson motorcycle, and two dozen lots in town. He built an A-frame house. He would take Marilyn shopping and buy her things she didn't even need. They joined the Quincy Free Methodist Church—top of the local society—and made friends there. Dennis and Marilyn loved the warmth, the music, the people at church, sitting in pews with the movers and shakers of Quincy. They wanted to have a family, raise it among the good Christian people, take vacations.

Dennis and Marilyn DeYoung didn't know how good they had it.

The early eighties were hard times to be a farmer in America—market prices fell to the dirt, inflation soared to the sky—but Dennis sailed through. His wheat crop was good in 1980, peas in 1981, sweet corn in 1982. Dennis's financial statements showed a net worth over $387,000. Not bad for a dyslexic thirty-year-old farmer. Dennis's land was worth three hundred thousand dollars, his mortgage was only one hundred thousand dollars, and his cash flow was as green as alfalfa. Marilyn and Dennis had their first child, Sara, in early 1983 as the frost was coming out of the ground. Dennis fixed up a yard full of farm equipment for the new season while Marilyn learned to breast-feed the baby. Dennis planted corn and wheat, and smiled through another good year: gross receipts over a quarter million dollars. The next spring, as Sara started to walk, Dennis planted peas, harvested them in midsummer and laid on buckwheat. Another nice crop and a happy fall. Shortly after the grass was swathed and laid in the field to cure, Marilyn gave birth to their second daughter, and they named her Heidi.

Dennis's net worth reached half a million—$509,528, to be exact—

according to a financial statement for the credit union. He had an excellent two to one ratio of assets to liabilities, $930,000 in assets to $421,000 in liabilities, to be exact. And he had a business plan to do $593,000 in sales and turn $95,000 in profits in 1985. No one doubted him. Dennis was farming his land and his father's land and rented land, more than a thousand acres in all, and he had three trucks and five hired hands. Other farmers were pleading for government help in the mid-1980s. Willie Nelson was holding the first Farm Aid concerts. Marilyn DeYoung was driving a year-old Buick Riviera and Dennis was thinking about buying more land.

It never happened.

The hay drying in the field at Heidi's birth marked the end. Dennis would not grow another profitable crop for ten years.

In 1985 his pea harvest was only one-tenth of what he had been expecting. The soil was too hard and dry. The sprinkler water seemed to flow off the hillside instead of running into the ground. Then the hay crop in fall produced a paltry four tons an acre. Dennis blamed the weather. He knew these things happened in farming, and he had learned to roll with adversity. On the dry lands, Dennis would suffer through a year without rain and the next year he would always bounce back. The rain would always come back. It was nature's law and more than anything farmers live with nature's law.

And, of course, they live with a lot of help—especially from the fertilizer company. In 1985, Dennis's unpaid bill for fertilizer, seed, and other farm goods from Cenex Supply and Marketing in Quincy went from zero at the start of the growing season to $50,912 in September. With his crops off, he couldn't keep up with the bills, and he didn't want to sell any assets for cash to pay the bills. Cenex had been giving Dennis a 7 percent discount for paying cash each month; now the company was charging him 18 percent interest. But that was business, and Dennis was confident he would bounce back, like so many farmers did.

It was not until years had passed that he looked back at the time his fortunes went bad and wondered about the fertilizer he'd bought in 1985.

He kept all his Cenex bills. They showed a big difference in fertilizer prices that could not be explained by ordinary market forces. Dennis had paid 9 cents a pound, 4.5 cents a pound, and 2.5 cents a pound for the nitrogen content of three different fertilizers. He didn't know where they came from or why they varied so. He trusted the Cenex field man to sell him good products.

David Nerpel, the Cenex salesman in that part of Grant County, had an aura about him from having attended a seminary. Instead of a priest, he became an adviser on farm chemicals, but he was unlike any other field man Dennis had met. Nerpel was always reading and learning, and his wife went to law school as an older student. In Nerpel, Dennis found a close friend. Dennis always enjoyed talking about religion, theology, and values. He tried to talk Nerpel out of abandoning Catholicism to join the Mormon church, as Nerpel told him the Cenex scientist Max Hammond, a Mormon bishop, was asking him to do. Marilyn DeYoung liked Nerpel's calm, in contrast to Dennis's excitability. She was glad Dennis had a pal like Dave Nerpel. Once the two men went to Seattle for a Mariners game but ended up spending all day goofing around the tourist district of Pioneer Square. Other times, they drove over to the coast for ocean fishing trips.

Nerpel was expert on the subjects of seeds and weeds and pests and fertilizers. *You'll bounce back next year,* he told Dennis at the end of the disappointing 1985 crop.

—

The soil took its first breath of spring air in 1986. Life stirred across the land in a canvas of browns and greens. The surface of the earth opened to the sun like a flower and scented the air. Crows were shrill on the high wires by the chemical companies lining the railroad tracks in Quincy, and a group of men was digging a hole in the ground.

The state had passed a law increasing corporate liability for dumping toxic wastes. Managers at Cenex Supply and Marketing decided they'd better stop pouring their surplus chemicals on vacant land, as had been the practice, and start collecting them for recycling or other disposal. The rinsate pond, as they called it, was lined with cement and looked like a big swimming pool. Cenex workers parked chemical-hauling trucks and tanks on a raised concrete pad jutting into the side of the pool and emptied the tanks.

The Cenex pond held seventy-two thousand gallons. It filled up fast; brown liquid reached within a hand of the top of the pond before the last frost. What was put in the pond? That was unclear. Some fertilizer. Some pesticides. Some unmarked cans of chemicals. Cenex didn't keep records of what went in the pond or what went out.

The pond was tucked along the railroad track in a thicket of ten-thousand-gallon tanks used to store fumigants and liquid fertilizer. Some people thought Cenex used the rinsate pond to mix a variety of leftover chemicals to sell as low-grade fertilizer.

One summer Cenex hired Len Smith to work in the yard. As the new man, he was given the dirty chores. Among them was emptying rusty old unmarked tins of unknown farm chemicals into the pond. Smith noticed the pond's depth fluctuated. Some nights it would be almost full to the top. The next morning, he'd be surprised to see the murky liquid had gone down a couple feet—roughly enough to fill a tanker truck. Where did it go? He did not know. He thought it was hauled away and spread on the land somewhere.

Len Smith never worried about it until he developed bone cancer some years later.

Cenex put a pump in the pond to spray the contents up in the air to speed evaporation. The fountain sprayed a shimmering mist of waste-water twenty feet high in the cobalt sky. It sprayed night and day. Sometimes the young people running around the track at the junior high school, a javelin throw away, smelled the chemicals pickling the desert air. Sometimes the teachers did, too, and complained about the smell.[2]

John Williams, Cenex manager, handled the complaints with a stoic manner that matched his solid build. He didn't think the smell was bad and he didn't think it would hurt anybody. Williams had been a Cenex man since he graduated from Quincy High School twenty years before.

The pond kept filling up. Len Smith was not the only one who said it went up and down, up and down, but Cenex had no records of any such thing, and Williams didn't blink when he denied it.

What was in the pond? Where did it go?

Williams answered simply:

Fertilizer.

Nowhere.

Williams never had to explain how a devil's brew of heavy metals—enough cadmium, beryllium, and chromium to qualify for a Superfund site—got into the fertilizer rinse pond.

———

Dennis broke even in 1986, but the next year, a windstorm wiped out his corn crop.

When times got tough, Cenex helped. The credit manager, Nick White; the Quincy branch manager, Williams; and Nerpel offered to help Dennis. They would give Dennis time to pay his bill, and deliver fertilizer on account. Dennis took more of the cheaper fertilizers because he couldn't afford the better ones.

The stacks of yellow invoices from Cenex piled up on Dennis's desk. He owed for seed, fertilizer, gasoline, nuts and bolts, work gloves, and $1.79 six-packs of soda pop. Marilyn used Cenex credit slips for fly sticks, rat poison, and cat food. Some months Dennis wouldn't even open the Cenex bill. He owed them twenty thousand dollars, then thirty thousand, then forty thousand. The October 1987 bill said, "Please Pay This Amount: $58,926.24" and had a smiley-face stamp in red ink and the words "Thank You We Appreciate Your Business."

In all, with the bills for seed, chemicals, water, labor, and equipment, the DeYoungs lost $337,000 in 1987.

Nerpel asked Dennis if he wanted to try growing potatoes. He could rent land in the area called Black Sands, a nearly perfect place for potatoes. The loose sand south of Quincy gave no resistance to tuber expansion. It was almost hydroponics, like growing plants suspended in a liquid mixture of the federal water from the Columbia River and liquid fertilizer from Cenex.

Dennis laughed, *You want me to grow for you in the Black Sands? I'm having enough trouble here.*

So he stayed on the farm. He'd pulled through before. But now it was just a slide that wouldn't stop.

Dennis and Marilyn moved out of the A-frame in town, which they rented, and into a cheaper, prefabricated home on his father's land.

Marilyn DeYoung gave birth to their first son, Seth, and Dennis got a letter he knew was bad news. The return address on the envelope was the Travelers Companies, Agricultural Division, Real Estate Investment Department. The company with the umbrella. Dennis opened it immediately. Travelers was calling in his mortgage unless he paid $12,335 in interest and $1,330 in property taxes right away. When Dennis couldn't pay, Travelers' local lawyer filed a lawsuit for foreclosure.

Marilyn's father, Richard Fenn, a successful farmer all his life, worked through three successive bad crops at this same time. Dennis tried to lend a hand by helping irrigate and pulling his combine in one of the fields.

When wheat grows well, the plants send out teller shoots and form big heads of grain. A farmer sitting on a combine looking down on a good crop of wheat sees nothing but the heads of grain. When the reel on front of the combine pulls the crop in to cut, he sees wave after wave of grain. The only time he might get a glimpse of the ground is when the reel splits the wheat, like parting thick hair. But when Dennis ran his combine on his father-in-law's field, he saw more dirt than heads of grain.

The land yielded just fifty bushels an acre. "For a farmer of his caliber on irrigated land," Dennis said, "that's junk."

So Marilyn's father lost their family farm in 1988. It was devastating. His life was farming. He blamed himself. It was the first time Marilyn saw her father cry.

Marilyn suffered a miscarriage.

Dennis's father was failing, too. Jake DeYoung had been successful for twenty years until the late eighties. Other farmers in the area were doing fine, and he didn't know what he was doing wrong.

Dennis held even in the 1988 crop year but made no profit, gained no ground on his debts. People started saying he was a bad farmer. Poverty will do that to you. His was the working slip-sliding life of a debtor.

Dennis, you don't have to farm, Nerpel finally said. *You've got too many abilities to do so many other things.*

Travelers, Cenex, the state, and the county were all pushing Dennis for money he owed. His lawyer said there was only one way to keep the creditors at bay. In January 1988, Dennis and Marilyn filed a petition with the U.S. Bankruptcy Court. While a Chapter 7 bankruptcy could have allowed them to wipe the debts clean, they filed under Chapter 13 instead, hoping to pay off all the debts in regular installments. The DeYoungs wanted to stand up, not run away.

Travelers, faced with the bankruptcy, modified the mortgage. If Dennis would pay $14,187 a year, Travelers wouldn't ask for the first payment until January 1, 1990, more than a year away. Dennis would have the 1989 crop year to recover. He agreed to the terms. Travelers withdrew the foreclosure and Dennis withdrew the bankruptcy petition and entered 1989 praying for a good crop to get back up on his feet.

Dennis raised wheat in 1989. But he saw too much dirt. He was expecting 125 bushels an acre but grew only 75. Then the price dropped from $4.00 to $2.35 a bushel. Dennis put the wheat in storage rather than sell it at a loss. A week before Thanksgiving, Marilyn had another son, named Lorin, but the DeYoungs were in no mood to celebrate. The state revenue department sued for unpaid business taxes. The IRS filed a lien. The Cenex bill said, "Please Pay This Amount: $73,682.73." And Dennis failed to make the payment to Travelers on New Year's Day 1990.

Soon a certified letter came. There was a picture of an umbrella on the envelope. The DeYoungs were in default on the farm loan again.

Dennis wanted to pull back, lease his hundred-acre circle to somebody who could make the mortgage payment, and move over to help his father on the family farm.

In the tenth year of their marriage, then, Marilyn DeYoung didn't slide over on the bench seat of the old Ford pickup to snuggle close while Dennis steered down the road. She sat and looked out the other window. Marilyn still drove the old Buick while other farmers' wives bought new vans and sport utility vehicles. Some of her friends from church said they felt sorry for her. She was raising four children all under seven years of age in a double-wide mobile home.

Marilyn told friends she would have left Dennis, not forever, but left him for a while at least, if she'd had enough money to afford to.

———

Cenex/Land O'Lakes served three hundred thousand ranchers and farmers in twenty-seven states. An unusually structured company, it combined the tax advantage of a farmer cooperative with the business dimension of a multinational conglomerate. The Cenex motto was "Where the customer is the company."

John Williams, the Quincy branch manager, had never known another business. Farm chemicals were his life. Born in Oregon, raised in Quincy, he had gone to work for Cenex directly out of high school, where, incidentally, he had known Dennis DeYoung. The dark, bulky Williams had an unflappable manner and sensible approach that people appreciated. The work of a field man was to be "a baby-sitter for the farmer," he said, and he didn't care if some farmers didn't like his way of putting it. After all, the metaphor placed the farmer in the position of a baby. But Williams was a straight talker, and he stuck up for his company over anyone.

By 1990, Cenex used new spreading equipment. It was customizable to lay down just enough chemicals each day and waste almost none. What little rinsing was done at the end of the day could be done safely in the fields. So the company decided to abandon the rinse pond in Quincy. Williams inherited the problem of what to do with the residue in the pond.

What a mess, he said.

Williams had thirty-eight thousand gallons of "what-all" to get rid of. "What-all" was a term for unknown, undetermined mixes of chemical. After most of the water evaporated from the pond, a brownish, gooey substance remained. It had the consistency of a milk-shake.

Sitting in his small office by the railroad track in downtown Quincy, Williams decided to simply spread the material on farmland as if it were fertilizer. He would say later he thought he knew what chemicals were in there, and that they were the same nutrients that were commonly applied to land all over the Columbia Basin.

But Williams didn't know what was in there, or if he did, he wasn't saying.

Williams took a sample to test for a few chemicals, including atrazine, a broadleaf weed killer. Atrazine was the nation's top-selling herbicide, used on 96 percent of corn crops, so it was logical to find it in the rinse pond. Williams didn't test for many other chemicals. The test was cheap, but perhaps he didn't really want to know that the discarded chemicals would have qualified as hazardous waste under federal law. Later, Cenex and the Environmental Protection Agency would find hazardous metals, cancer-causing chemicals, even some radio-active materials, along with fertilizers and pesticides in the residue of the rinse pond. The "what-all" was a witches' brew.

Cenex and John Williams faced a choice. They could store the contents of the waste pond at the Northwest's only plastic-lined and licensed hazardous-waste dump, in Arlington, Oregon. "You own everything you put in Arlington for the rest of your life," Williams said. "It's not a very good option as far as I'm concerned."

Or they could spread it around as fertilizer.

Larry and Julie Schaapman knew Dennis and Marilyn DeYoung from church. Larry was one of the church leaders. They had a small family and a successful farm out on Road P NW. The DeYoungs admired the Schaapmans.

Larry also had an unusually close relationship with Cenex. One day in early 1990, a Cenex man told Schaapman the company had a great deal for him. Cenex would pay all of Larry's costs and a custom farming fee—guaranteed profits—if Larry would lease a certain plot of land and let Cenex put the material from the rinse pond there.

Dennis DeYoung's land.

The field man told Larry the "product" from the rinse pond was just fertilizer and atrazine. Larry checked with John Williams. Larry trusted Williams, so it was easy to accept the deal from Cenex. Larry knew Dennis needed money, so it was easy to lease Dennis's land.

That was how John Williams put together the plan, the land, and the company farmer to take care of the Cenex problem. Williams wrote the Cenex regional supervisor.

Re: Plan to clean up wash water pond at Quincy

We have approximately 38,000 gallons of product to dispose of. If we haul this liquid to Arlington's Chemical Disposal we will be required to dry it down to dry matter. Arlington will not accept liquid product. We would add a drying agent to it which would increase the weight two times. This would result in 500 tons of disposable product.

The cost of disposing at Arlington is about $280.00 per ton, add $20.00 per ton freight and 250 ton of drying agent at $80.00 per ton which equals a total of $170,000.

Also keep in mind that whomever uses Arlington also owns a percent of it if any contamination occurs or they need funds to clean up the area.

We have had this pond tested for chemical content and feel we can spread it on corn ground without enough carryover to affect any crops the following year.

We can rent 100 acres of corn ground for $125.00 per acre plus power and water. Total operating expenses including rent would be $300.00 per acre. Approximate return would be $400.00 per acre or a total profit of $10,000.

I am anxious to make a decision on this matter as soon as possible.
Sincerely,
John Williams
Branch Manager

And so they decided to save the $170,000 disposal cost and take the $10,000 profit instead. They would simply call the waste a product and spread it on Dennis's land.

Let nature take its course. The corn could use the fertilizers and atrazine. The land would act as a natural filter for the hazardous chemicals hidden in the mix. The chemicals would wash through the land. Or so they hoped. The cost saving was also a boon to Williams and other managers who had a profit-sharing plan at the Cenex branch in Quincy.

Years later, Williams would add a new twist to the story. Williams would say he'd talked with a state official about what to do with the waste pond and claimed he was told it would be fine, just fine, to spread it on farmland. Williams would remember very specifically talking with this official out in front of his office in Quincy, but he would say, under oath, that he couldn't remember the person's identity.

Nothing about that conversation was mentioned in the letter to his own boss.

Williams was backed up in this unlikely assertion by Ron Kopczynski, his protégé, who had replaced him as head of the Cenex field men in the Quincy area. Kopczynski would say a state official also told him in a phone call it was okay to dump the waste on a farm.

"Somebody, and I can't recall the guy's name, who he was, but I asked him what he felt the best way to dispose of this, and I gave him what we had in here. He at that point said to me that the best way of disposing of it would be to apply it to the ground to be used as fertilizer, but that our discussion would be gone. If anybody ever asked him, he would say he had never talked to me."

Nick White, the Cenex regional credit manager, also said he happened to be in the Quincy office and heard the conversation the day Kopczynski called the capital. Alas, White could not remember the name of the state official, either.

But Cenex got the green light on the deal. So they claimed.

In a lawsuit, years later, Williams would also claim that Dennis had wanted the waste to be put on his land.

"Well, Dennis and I was sitting in the front of the office one morning and he was talking about needing a renter for that piece of property to make his payment. We come to a mutual agreement that we would rent it and apply the rinsate pond to it, very high in fertilizer, get some benefits out of that. He would make his mortgage payment. We'd have a place to get rid of this."

Both Dennis and Schaapman denied John Williams's version of events. The truth seems to be that Cenex kept Dennis in the dark. And no one knew exactly what was really in the semisolid waste they called a product. No one checked; no one asked.

On a Monday in early March of 1990, Schaapman asked Dennis to come by his office and talk about the land. They'd seen each other at church the day before. Deal time. Larry drafted a simple lease agreeing to pay Dennis $12,500 for the use of his hundred-acre circle for one year. Dennis took the money. It was just enough to pay the mortgage.

Less than three weeks later, as the soils were getting warm and dry enough to work, Larry signed a side deal with John Williams at Cenex. Cenex agreed to pay Larry $125 an acre rent on Dennis's land, plus $10 an acre to plow the ground, $12 an acre to plant corn, $7 an acre to harrow or level the ground, and $25 an acre to manage the irrigation over the summer. Cenex would apply all fertilizer, chemicals, and seed for free. Cenex would help sell the corn that grew. All of Larry's expenses would be taken out of the profits, and the remainder, if any, would be split between Larry and Cenex, fifty-fifty.

Larry penciled it out. The deal would give him twenty-nine thousand dollars, maybe more, risk-free. Larry took the deal.

Quickly in the days that followed, Larry plowed the field, Cenex workers sucked the mix out of the waste pond with a pump and hose, and truck after truck delivered it to Dennis's field.

Dane Lindemeier worked as a Cenex applicator. He sat in the driver's cab of a big yellow Terra-Gator, a vehicle supported by three wheels so wide they almost floated over the soft spots of the field, like snowshoes. The flotation wheels kept the soil loose for the chemicals to go down.

Dane hauled a two-thousand-gallon stainless steel tank on the back of his Terra-Gator full of waste pond residue. He had watched workers mix up the pond to suspend the solids and pump it out of the pond into his tank. The soup was light green, the color of a fertilizer. Dane was told the mix held the pesticides atrazine and trifluralin. He saw right away that that didn't make sense. The atrazine would protect corn but kill beans, and the trifluralin would protect beans but kill corn. He was used to spreading fertilizers or pesticides, not this.

"I told them I didn't want to spread the stuff," Dane said later.

Spread it, Williams ordered.

Lindemeier obeyed.

"I'm not a union man, I'll put it that way. I've got a family to feed," he said later.

Load after load was delivered from waste pond to farmland. It was put on smoothly, professionally, a chemical blanket laid lightly on the earth. The Terra-Gator covered Dennis's land one way, then the other, crisscrossing, flood-jets calibrated to spray three hundred gallons to the acre.

"It was not dumped," Williams said later. "It was spread on there."

Dennis, the owner, was nowhere to be seen.

A short time later, Cenex hired a man with a Caterpillar to break up the six-inch concrete walls of the rinsate pond and cover it over with dirt. The waste pond disappeared. Nobody ever ran a full set of chemical tests on the Cenex "product."

Dennis didn't see the material on his field until some days later, when Larry called him over to fix the sprinklers. It looked like a heavy coat of fertilizer, grayish on top, chalky in places, covering the field. Larry didn't tell him any more about it, and Dennis didn't ask.

Larry started watering it in. Dennis said Larry applied vast quantities of water in the spring and summer. Larry denied that. The circle was on a bench of land above a small lake that drained into the Columbia River.

Larry planted corn. The leaves grew purple, the stalks were stunted, and the roots were unnaturally bare. Hundreds of feet of field were simply dead. The stalks that survived were four inches tall.

Dennis talked to Williams because he knew Cenex was helping Larry.

The corn is dying out there.

It's the cold weather, Williams replied.

Well then, why don't you just get out and replant it?

It will come through okay, Williams said.

Dennis got the idea that Williams thought he was wasting Williams's time. The Cenex manager had a lot of business to do. He had a lot of farmers to take care of.

As some weeks passed, the big tires that held up the sprinkler pipes dug ditches in their slow circle around Dennis's field. The gearboxes dragged in mud, and one day the whole apparatus caught and flipped. Larry called Dennis, yelling. Dennis worked hard to get the circle back up and put on a new pump. He didn't want to be blamed for a poor crop. Larry said the corn would come through all right. It never did.

Dennis was busy running more than four hundred acres on his father's farm, and they weren't growing very well, either, though Dennis had irrigated the field before planting, tilled it, planted on time, and used all the chemicals Cenex recommended. The field man Nerpel came over one day while Dennis was working on some planting equipment, and they started to talk about the usual subjects—the weather and what was wrong with Dennis's crops. This day Nerpel acted especially worried.

And what he told Dennis—just between the two of them, standing alone in the garage—Dennis would never forget.

You oughta check into what was in that fertilizer, the field man said.

Then they got talking, really talking. They went inside. Marilyn came home from shopping and saw Dennis in the rocking chair in the house, holding one of the babies, talking with their good friend Nerpel. Marilyn was tired of hearing Dennis whine about what probably happened, might have happened, could have happened. But now she heard some of the things they were saying about the fertilizer. Some things started to add up.

Nerpel was calm.

Dennis was rocking.

Marilyn was starting to work up a red-hot anger.

The next day, Dennis drove to the Cenex office in Quincy and walked into John Williams's office. Dennis looked Williams in the eye and asked what was in the fertilizer. And again he was told not to worry because it was all tested and fine.

The corn crop yielded one and one-half tons an acre, one-third of the normal amount. Larry blamed the weather and Dennis's sprinklers, but it turned out Cenex was to blame. A test would show the Cenex product had illegal amounts of the herbicides Avadex, Treflan, and metribuzin, and the corn died from chemical overdose.

It was marketed and sold anyway, as animal feed. To this day, no one has traced where the poisoned corn from Dennis's land was sold.

At about the same time, most of the 350 acres of beans at his father's place died, too. Dennis harvested a pitiful nine bags an acre and sold the whole crop for thirty-six thousand dollars, worse than bad. And it was either his fault or somebody else's. Seeing the Cenex connection to both crops made Dennis and Marilyn think Cenex had set them up for a fall.

There was another reason for that suspicion, in addition to Nerpel's warning. Marilyn's brother worked for the city of Quincy at the settling ponds used for Lamb Weston and Simplot food-processing waste. One day they were cleaning out sludge from the settling ponds, vegetable peels and runoff from the food production lines that ought to make an organic soil additive. The pond operator asked Marilyn's brother if he knew anybody who could use it. He suggested Dennis. Some men— Dennis said he can't remember exactly who—came out and told Dennis they'd pay him one hundred dollars an acre to put it on his father's land after the bean harvest in the fall, and it would improve the soil. They'd just have to test the field first.

But when they conducted the tests, they changed their minds. Dennis recalled an expression of fright or concern in something one of the men said about his father's land:

We don't want any part of that.

It occurred to Dennis later that they didn't want it because his father's land was already contaminated, like his own. He had no proof for that assertion.

After the beans died on his father's land, after he confronted John Williams, Dennis saw Dave Nerpel no more. Dennis blamed Cenex for fertilizing with a cheap, low-grade material he'd never heard of. Nerpel, publicly at least, blamed Dennis for not following his advice. A short time later, Nerpel moved a hundred miles away.

By now, Dennis was indebted to Cenex for $150,000; state and federal taxes and employment insurance were long overdue; and Travelers was breathing down his neck like fire. So once again Dennis drove to his attorney's office on Hemlock Street in nearby Othello and sat down to talk. There he learned for the first time that his attorney was also working cases for Cenex. Dennis thought that was a conflict of interest.

So the attorney withdrew as Dennis's lawyer. Dennis always remembered his words: *I have to go with Cenex.*

Through Thanksgiving and Christmas, the hundred-acre circle of land lay in repose. Through a bitter winter, the soil was dormant, almost dead, enduring until it was touched by the breath of spring.

—

Dennis did not want the land back when the lease ended in early 1991. He would not take the land back. He didn't know what Cenex had spread on the circle. He wondered if the chemicals that killed the corn would seep into groundwater or poison well water.

We could even be liable as a hazardous-waste site, he told Marilyn. *This is bad stuff. We can't take the responsibility for it.*

Dennis talked to John Williams, but Williams kept saying the land would be fine; the problem was the weather, the sprinklers. Dennis told Williams Cenex should just buy the land. He'd get out of his debts and they could deal with the dirt. He'd lose the farm but win a fresh start.

Dennis was not always nice about asking Cenex to buy him out. He threatened them. He told Williams to buy him out or he'd get a new lawyer and sue.

Williams talked to the Cenex credit manager, and they agreed the company ought to buy Dennis's land, but they had to get approval from headquarters in Minneapolis first. There, a Cenex vice president said the land had too high a mortgage. Cenex was in the business of selling to farmers, not owning and farming their land. So Williams went back to Dennis and told him they'd tried to buy the land, but couldn't.

Good luck, Williams said. *The land will come back. It will be fine.*

Dennis didn't believe him. He came up with a simple way of refusing to take responsibility for the land: He refused to turn the water on. That way he could not be blamed for what grew there or what washed off the land.

So it was a stalemate until the earth took its tentative breaths of air after the last hard frost in March, and then Williams agreed to a deal with Dennis. Cenex itself would lease the crop circle and the orchard from Dennis for a year. Cenex would pay $46,524 to Dennis's account at Travelers to make up two years of missed payments; and Dennis and Marilyn DeYoung would give Cenex a $130,000 mortgage on the land to secure the debt they already owed from all those yellow invoices in the late 1980s. All told, the deal would leave Dennis's farm with $230,000 in debt, a heavy debt, but the land was worth easily $300,000, maybe $400,000, so nobody would be standing naked on the deal. Dennis thought it was a fair payment for him to receive in return for a year's worth of crop rights, including the apple orchard. If the land bounced back, he'd be fine, and meanwhile he would drive his Mack truck for hire and work his father's land.

Cenex paid Dennis and Marilyn one dollar to seal the deal. Dennis and John Williams shook hands.

Travelers dismissed the foreclosure action as soon as it got the money from Cenex.

A little while later something suspicious happened again, to Dennis's way of thinking. The credit manager, Nick White, phoned Dennis and asked him to come into the Cenex office in Quincy. He said there was a paper he had to sign to turn the water on. Dennis drove his Ford into town and parked by the tracks and walked into the little Cenex office. White greeted him with an eager-to-please look like they were friends, and John Williams joined them in a less dark mood than Dennis had seen lately.

Where is your wife? White asked.

Well, she's helping over at the school, Dennis said. *Why? Do you need my wife here?*

We need her to sign this quitclaim deed with you to turn the water on, White said, according to Dennis's recollection later.

Now wait a minute, Dennis said.

Why would Cenex need a deed on his land to turn the water on? The deed was signing the land over to Cenex. He had already signed a mortgage and promissory note and he didn't want to sign a quitclaim deed.

You can go ahead and sign it and nobody has to see it, White assured him, by Dennis's account. *We can just put it in the drawer.*

If I sign a deed, I'd want the whole world to see it. I would not want it put in a drawer, Dennis replied.

White and Williams never did produce a document for him to sign. Dennis left a short while later and drove back home.

Something is wrong, he told Marilyn.

He wished he had a lawyer.

And when he next went to the Cenex office to talk about the water, they had dropped the idea of a quitclaim deed. John Williams just had Dennis sign approval for the Columbia Basin Project to turn on the water for the season.

As far as anyone knew, the DeYoung property marked the first time that Cenex/Land O'Lakes had gone into direct farming, even on a lease. In April, the crew hired by Cenex planted Sudan grass. Normally you'd plant a better cash crop in the spring and follow up with the lower-value grass for sale as seed or hay in the late summer. But Williams said Sudan grass was the only crop they could plant on Dennis's land because of the late start.

Sudan grass is known as a good accumulator of heavy metals from the soil, a biological sponge.

Dennis figured Cenex was trying to cleanse the soil. The water soaked the chemicals down. The grass sucked the chemicals up.

A lot of plants don't show the potentially toxic chemicals they absorb. Some of the most hazardous chemicals, such as cadmium, are poisonous to animals but harmless to plants. Others are phytotoxic, or deadly to plants, before they would ever accumulate to a level that would hurt animals that ate them.

The Sudan grass, normally easy to grow, was a mess. It looked burned and didn't grow at all in some strips of bare dirt, and grew green in others.

Dennis took pictures. He talked to John Williams.

You've ruined my land, he said.

Williams replied the land wasn't damaged, but he admitted Cenex wasn't managing it right.

We did a poor job.

You sure did.

Williams told Dennis he was ungrateful. Why attack the only people who would give him money when nobody else would lend him a dollar?

One day in June, Dennis and his father drove over to Nick White's office again. They sat down and Dennis sighed.

We've got to talk.

White looked genuinely puzzled. *Talk about what?*

The rinse pond. The rinse pond has ruined my farm, Dennis said.

Now White looked not only concerned, but dumbfounded. Later he would say that he had been trying to work the money problem through in his mind but had not known he'd be asked about chemical damage to the land. "This one caught me off guard. At that time it became a big issue," he said.

Some days later White offered Dennis $300,000 for the property.

Too low, Dennis said. *It's worth $560,000.*

Too high, White said.

Dennis asked him how much Cenex would have to pay if it was a Superfund site. White said that would never happen.

Dennis found a new lawyer, a Seattle lawyer named Bradley Jones, who drove over to Quincy just after the Fourth of July and dictated this description of Dennis's field:

"Despite good weather and regular and massive watering by Cenex, nothing has appeared except stunted, patchy growth. The grass should be more than two feet high, and it should cover the ground completely. There are large bare expanses of ground, indicating soil sterility."

The next day Dennis and the lawyer started filing complaints with every responsible agency they could think of: the state department of agriculture, the state department of ecology, and the U.S. Environmental Protection Agency.

They said Cenex, the friend of the farmer, had poisoned this farmer's land with a "product," a "fertilizer," that was really a toxic waste.

Steve George of the state agriculture department pulled the complaint to investigate. His first thought was that Cenex might have broken the law on pesticides. On a sunny Thursday morning, George drove his Dodge pickup to Quincy. He spent all day looking at Dennis's land and talking to the unflappable John Williams and other people at the Cenex office in Quincy. It was the first official investigation.

Williams and White didn't like it one bit.

The next day, Cenex hit back at the DeYoungs as hard as it could with a letter exercising the accelerated payment clause of the promissory note Dennis and Marilyn had signed; in simpler terms, calling in the IOU. White knew the DeYoungs were broke. *Tough.* White gave them two weeks to come up with $130,000 in cash.

The battle joined, Dennis's lawyer hired a soil chemistry expert from Washington State University to inspect the land.

And Dennis brought back the real estate agent who had sold him the farm a dozen years earlier. *How much is it worth now?* Dennis asked. Jim Weitzel was a former air force captain, aerospace company manager, and Grant County commissioner. He had also served as chairman of the county hay association and grown plenty of Sudan grass in twenty years of farming his own land. Weitzel knew bad dirt when he saw it. He saw it on Dennis's land. There were strips fifteen feet wide where not even a weed would grow.

Under normal circumstances, Weitzel calculated, Dennis's farm would be worth $429,234. His appraisal praised the "well producing orchard" and buried drainage line that Dennis had put on the land. "The area is one of the most stable and desirable farm areas in the Basin, with few sales," Weitzel wrote. "Many qualified buyers are normally looking for ground in the area."

But with this chemical damage, he said, it was worth next to nothing. Weitzel said he would refuse to sell it at all.

The liabilities are too great, Weitzel said as he walked away from Dennis in the field, leaving him shaking his head.

The lawyer wrote a letter accusing Cenex of using Dennis's farm as an unregistered hazardous-waste dump. The waste pit was so foul, Jones said, that when it was sprayed in the air for evaporation the railroad complained about fumes corroding equipment nearby and affecting workers on passing trains. It was so toxic that even after a year on Dennis's field, a simple crop for hay wouldn't grow. Jones said an "extremely rank stand" of Sudan grass covered only 22 percent of the ground.

Shortly after he started filing complaints, Jones added, Cenex poured expensive fertilizer on the grass, then cut it and wrapped it into bales to send away. "The Sudan grass was withering at the time it was mowed down in spite of triple the normal water schedule, large applications of nitrogen to 'green up' the grass, and abundant good weather."

It made no sense financially to raise Sudan grass for hay as the sole crop in a growing year, or to pour expensive nitrogen on it—except to try to extract the toxics from the soil.

The lawyer threatened to seek triple damages under a state consumer protection law and federal action with a federal racketeering law aimed at the Mafia but expandable to emcompass interstate business conspiracies.

Dennis's lawyer tried to settle the case early by making an offer in one of his first letters to Cenex. He'd drop the case if Cenex paid $430,000 to Dennis for the land; released the $130,000 promissory note and mortgage; changed the accounting statement to show the Cenex payment had been a lease, not a loan; and disavowed the illegal actions by its local personnel.

Cenex's own lawyer was a former Forest Service smoke jumper named Michael Rex Tabler with an office by the Grant County Courthouse. Tabler had a flattop haircut that fit his taut, controlled manner. He thought about the demand letter from the Seattle lawyer. The money wasn't a problem, but the precedent was. And it was out of the question to disavow the Cenex manager Williams. Out of the question to admit he'd done anything illegal.

The answer, Tabler decided right away, was no. And Dennis better get the $130,000 paid or he'd be sued.

Larry Schaapman worried about the lawyers and investigations and the talk about racketeering and hazardous-waste dumps. He could easily be dragged in. Williams assured Schaapman that Cenex would pay all of his legal costs and any settlement if there was one. Larry wouldn't have to risk a thing. He would stay a full member of the Cenex team.

The late-summer heat thinned the air and baked the earth. Farmers ran the water in the evening to reduce evaporation. Stars radiated in the desert sky at night.

One day Steve George and Viki Leuba sat down for coffee at Bob's Restaurant in the town of Moses Lake, south of Quincy, to talk about Cenex and Dennis. Leuba had a complicated job as head of the hazardous-waste and toxic reduction program in the state ecology department. On the one hand, she helped companies recycle hazardous wastes, and on the other, she was their watchdog.

George, from the state agriculture department, had already been over to Quincy, seen the Cenex operation, talked with John Williams.

He'd been told Cenex applied a Dow Chemical Company product called Curtail to control thistles on Dennis's land. Curtail was a 32 percent solution of 2, 4-D, the herbicide made infamous by Rachel Carson's *Silent Spring*. Curtail was legal and widely used, but the 2, 4-D meant it would be illegal to harvest the Sudan grass on the land for sale.

No problem, Williams responded, according to Steve George's meticulous notes.

"It'll be treated as fallow land," the Cenex manager said. "We won't take any crop off it."

So the state employees both felt assured as they drank their coffee at Bob's Restaurant. They had always trusted Cenex.

OCTOBER 1990 — A FEW MILES WEST OF QUINCY

Ruthann Keith's Appaloosas were many things to her: money, work, love, pride. She loved to talk about them while she sprawled in a lawn chair in the shade outside her ranch house.

Many of Ruthann's sixty horses came from a stallion descended from the Ghost Wind horses of the great Nez Percé, one of the last Indian tribes to fight back. A white-haired native passed down the Ghost Wind story in the 1860s. He said he had witnessed a battle between the Nez Percé and the United States Army. The Indians were trying to regain land the government had ceded to them, then taken back when someone struck gold. The man said he witnessed the end of the battle over "Cheater's Treaty" from afar, saw the army cavalry slaughtering unarmed Indians. And the cavalry saw him. They raced after the lone Indian. He thought they were after his beautiful stallion. The horse outpaced the cavalry; they eluded patrols and spent the winter hiding in a cave on the Montana-Canada border. They lived many years longer, wandering the West, the horse admired in several states. Years later, after he told the story, people excavated the cave and pieced together the movements and lineage of the colts fathered by the Appaloosa stallion.

Primitive horse species native to the Americas became extinct eight thousand years ago. Why they died out is one of the great mysteries of science. Most historians believe Spanish explorers brought the first horses back to North America. The Spanish ran horse farms in Jamaica by 1515 to supply Francisco Pizarro's expedition to Peru, and in Cuba for Hernán Cortés's invasion of Mexico in 1519. The horses carried Spanish explorers and their gear north, south, and west through the New World.

Ruthann had a different theory for the ultimate origin of some of her own horses. She thought Russian fur traders brought them to the West Coast. That's why, she said, coastal Indians were riding horses before the Spanish explorers arrived. The Ghost Wind horse, in Ruthann's telling, probably derived from the Russians.

Ruthann started with the champions when she started seriously breeding Appaloosas. Other breeders had crossbred a lot of quarter horse and thoroughbred; pretty soon their line was one-thirty-second Appaloosa or one-sixty-fourth Appaloosa. Ruthann avoided crossbreeds and looked for true sons and daughters of the earlier Appaloosas.

"I went back and got the oldest stuff and just started over," she said.

One of Ruthann's stallions was the last one kept by Claude Thompson, founder of the Appaloosa Horse Club. When Thompson started breeding Appaloosas in the 1930s, there were only 325 of them left in the world.

"They're not just horses. They're endangered bloodlines," Ruthann often said.

Royal Sunshine was the first registered Appaloosa Ruthann bought. The colt's mother, Gypsum Bow, came along, too. Gypsum Bow, whom Ruthann called "Ma," was the horse who sold her on Appaloosas. There was nothing Ma couldn't do.

Appaloosas have a distinctive patterned spotting, and Ma wasn't real pretty. Her head, especially, had blotches of color that other people thought made her an ugly horse, but Ruthann always said you don't ride the head. Luckily, Ma's head was genetically recessive; none of her heirs had a head like that.

Ma was lame from an injury when Ruthann first brought her home. It took a year to heal up. She started working cows. The old horse was Ruthann's best mount. Appaloosas are relatively light horses, weighing about one thousand pounds. One day Ruthann and Ma were corralling a sixteen-hundred-pound part-Brahma bull injured giving birth. The angry Brahma turned on Gypsum Bow, hit the mare under the rib cage, just behind the saddle, and started to lift.

"I thought she was going to flip her over," Ruthann said. "I don't know why she didn't. But we scooted out of the way and we got the cow all doctored, and I rode the mare all the rest of the day, and I pulled the saddle off and she had three broken ribs up underneath the saddle. I didn't know what to do about broken ribs on a horse, so I just kind of put my hand up there, and, you know, the rib cage comes straight down at that point, and I just shoved down and popped 'em back into place and brought her home and laid her off for a few weeks.

"Then I went back to riding her again. She could do anything."

When Ruthann had a bad day roping, Ma would put her on the calf and keep her there until she got the rope over. She just kept going. She was sleek and strong and broad of chest. She had a lot of heart.

As she thought back, Ruthann remembered wondering why they were watering Dennis's field so much that summer. The sprinklers ran so heavy that a stream ran down the hill. That had never happened before. Ruthann knew the land well. Her grandfather had home-steaded the hundred acres Dennis DeYoung now owned. Ruthann had ridden around the fields since she got her first motorcycle at the age of ten. She was still riding a motorcycle in her forties, kick-starting the beat-up old bike and gunning down the dirt roads, sometimes with a dog on her lap. Ruthann had watched the water run off the field. Such a waste of water. She thought the people tending the field for Dennis were incompetent farmers.

One day in the fall, Ruthann stopped in at the Cenex Supply and Marketing farm store on A Street. She said she was looking for some grass to mix with her alfalfa and oats as a transition winter feed for her horses coming in from pasture. Pure alfalfa would have too much pro-tein and calcium, especially for horses just one or two generations off the wild grass of the plains.

You know, there's some grass hay not very far from you that Cenex has, said Janet Jenkins, who was working the front counter.

Really? Where?

That circle out there.

Circle?

Yeah, you know, out there on Road 11.

Oh—you mean DeYoung's circle?

I guess Cenex is farming it now, Jenkins said. *It would probably be just what you wanted. It's Sudan grass. I don't think it's real expensive.*

Well, I've never fed any Sudan grass.

After it's dry it's supposed to be fine. Maybe you should call John Williams.

Ruthann stopped by the circle to look at the hay. It smelled like fresh-mown grass. The long stems were cut and drying. Larry Schaapman had harvested 5,422 bales of Sudan grass and billed Cenex for his work. Ruthann didn't know the land had produced less than a normal healthy yield in its second year after the Cenex waste pond was spread there; she didn't know about any inspectors' visits or any promises not to take crop off the land. Almost nobody knew.

Ruthann called the county extension service. She was told mixing in Sudan grass for her feed was acceptable practice. She called Williams and ordered forty tons. He could sell her thirty tons at $40.00 each, or about $1.50 a bale, he replied. A good price and close to her farm.

It's a deal.

From time to time in November, Ruthann drove her pickup truck the mile and a half to Dennis's circle. She'd toss thirty or thirty-five bales in the bed of the truck, haul them back to her farm, and stack them in piles. The bales weighed at least seventy pounds each. Ruthann liked hard physical work.

All summer, her horses had grazed on orchard grass, rye grass, clover, alfalfa, and dandelions in Ruthann's pastures, first one field then another, rotating to where the grass had grown. As the angle of the sunlight marked another late autumn, it was time for the horses to come in to eat.

A lot of Ruthann Keith's Appaloosas were first generation off the range. They had never been removed from the native grasses that covered the area before the white man came. Ruthann treated them very well.

She started feeding a little bit of Cenex's Sudan grass in November to wean the horses off the pasture. Then she mixed the Sudan grass with the rich alfalfa. The grass perfumed the air. She thought her Appaloosas would love it. They did.

"They scarfed it right down," Ruthann said.

Some of the horses started losing weight. Some of them started suffering from diarrhea. Ruthann thought it was colic or too much calcium from the alfalfa. Then Sunshine got sick, and by December, Gypsum Bow was losing weight, too. Ruthann wondered what she was doing wrong.

Maybe it was parasites, but after she wormed the horses, they looked worse. She didn't suspect the Sudan grass hay; in fact, she fed them more to try to fatten them up. They just got sicker. Lesions developed on their long necks and broad shoulders.

It was odd. Ruthann had four groups of horses. She fed the Cenex hay to only two of the groups, and only four of those eight horses got sick. Ruthann looked for clues within Gypsum Bow's family lines. Maybe allergies were causing the lesions on Ma and Sunshine, she thought, because the illness seemed to run in the bloodline, and allergies can be genetic.

In January, Ruthann finally noticed that not all the horses in a group ate the same food. She thought she figured out the problem. It was, she believed, a combination of bloodlines and bad hay.

Dominance is a family trait among horses, as cousins and aunts hang out together and even babysit for each other because of their similar temperaments. The feeding behavior of Ruthann's horses followed bloodlines. The more dominant feeders in the two groups were eating the alfalfa and leaving the less desirable grass hay for the others. It was like eating the candy first, then leaving the broccoli. Ma and Sunshine got what was left, often eating off the same bale of Sudan grass hay.

Sunshine was the first one to get desperately sick. Ruthann put Sunshine on some bran and banamine. She got a little better. Then she got a lot worse. One day the horse strained so hard to move her bowels, she ruptured, and her intestines spilled out on the barn floor.

Ruthann picked up the six-foot large intestine as best she could, put Sunshine in the horse trailer, and rushed her to a veterinarian in Wenatchee. It was a half-hour drive at highway speeds, and Ruthann couldn't go faster pulling the trailer. She was hoping the horse would somehow survive the ride, but she imagined Sunshine crumpled on the trailer floor as she steered north up the Columbia River, past orchards, wheat fields, and gas stations. Ruthann wasn't one to panic, but her own gut hurt. Sympathy pain.

The vet had never seen anything like it. He said she would not recover and all he could do was put her down by lethal injection. Sunshine was only twelve years old.

The day before Sunshine's death, Ma had started showing signs of distress. Ruthann checked her again as soon as she got back home from the veterinarian, empty trailer in tow. Ma was worse. The old horse worried and fretted and started straining like Sunshine had. The next morning, after one of the worst nights of her life, Ruthann knew she couldn't let Ma suffer like Sunshine had.

She led the mare to the place out back where horses were buried on the farm, crying all the way. She could hardly see to open the gate on the fence and to put a bullet in the gun.

"I had to take her out there and I stood her up and she kind of looked at me and I petted her and she looked at me in the eye and then she kind of dropped her head and closed her eyes like this, and it was just the perfect shot. It was like, 'Okay, I know you gotta do this,' and she just made it easier for me."

Ruthann had owned Ma for fourteen years. Nobody else could ride her. She grieved. She still didn't know exactly what was going wrong.

She figured it out after her Bear Paw line got sick.

A horse named Missie was the next to get desperately ill. She was about twenty years old and pregnant, one of the last producing daughters of a national champion named Bear Paw. Ruthann started popping Missie with banamine and mash and forcing fluids, but the mare lay down and stayed there. Horses and cows can lie on the ground to take a nap, but if they stay down too long, their digestive organs stop working. Ruthann and her husband, David, had to push and thump Missie to get her up. It took two or three weeks before Missie started to perk up on her own, but then she miscarried her foal.

Another mare, Requesta, developed sores and a tumor on her jaw, and she, too, miscarried. It took two seasons before Requesta got back in foal, and that one was stillborn. The one after that was finally born alive. Requesta was one of the last daughters of Simcoe Snowy Rock, a national champion and one of the top producing stallions of all time.

In January, Ruthann stopped feeding the Sudan grass hay to her horses. A month or so later, she was in town one day to pick up stamps at the post office. She saw Dennis DeYoung getting into his rig.

Hey Dennis! Ruthann hollered.

He looked up and walked over.

Dennis said, *I understand you bought some of that Cenex hay.*

Yeah.

Well, how'd you like it?

Well, I didn't. *I didn't like it at all.*

Ruthann told Dennis how her horses had sickened and died.

Oh, man, Dennis said. *I knew I should have called you last fall.*

Ruthann was puzzled. *Why should you have called?* she asked.

Well, we think they dumped something on that circle out there.

Dumped something. Dennis said he didn't know what. The field man Dave Nerpel had warned him. Cenex was calling it fertilizer, but the chemicals came out of the waste pit.

Ruthann went home and called her veterinarian. He'd never dealt with reactions to toxic chemicals; he suggested she call the state health department. Somebody there suggested she call the EPA, where somebody else put her in touch with a toxicologist, an expert in poisons.

Ruthann described the weight loss, diarrhea, sores, and organ rupture. She said the toxicologist told her they were classic symptoms. She asked where she could send the hay to be checked. He told her a tox screen would cost about a thousand dollars a horse unless she knew what she was looking for.

Well, I got sixty head of horses out here. There's no way I got sixty thousand dollars to find out what's in those horses.

Ruthann never tested the hay. When her mares started dropping healthy colts again, she was thrilled and relieved. But then after a couple weeks of nursing, the colts started losing hair in little patches on their chests, necks, and shoulders. She worried, but they recovered their hair. Ruthann's Appaloosa studs had never been fed the grass and never developed any symptoms.

Later that spring, Ruthann noticed something about the yearlings that had eaten the hay but not fallen as sick as Sunshine, Ma, Missie, and Requesta. It was their hooves. Horse hooves are made of the same kind of material as human fingernails and grow about a third of an inch a month. Changes in the hooves mark major events in the health of a horse, like tree rings mark fires and droughts in a forest. Ruthann had seen discolored rings come down hooves of horses in the months after they were hauled two thousand miles. The yearlings' hooves were growing smooth, then suddenly there were a lot of ridges. The time coincided with the feeding of the Cenex fertilized hay.

I don't know what's going to happen with these babies, Ruthann told her husband.

She ended up giving them away or selling them cheap and telling the buyers they'd eaten something bad. She kept a couple yearlings, but they never muscled out like the others. They were stunted and skinny. She wouldn't sell to breeders.

Ruthann called the state department of agriculture. After a delay, a worker came out, put on double-thick plastic gloves up to his shoulder, and took a small sample of the hay. (Ruthann didn't think he took nearly enough.) After another long delay, the report came back to Ruthann saying they found the herbicide atrazine in the hay. That wasn't a huge surprise because that is what Cenex told the inspector was in the fertilizer after Dennis complained. They didn't test for heavy metals.

Ruthann tried to talk to Cenex. John Williams in Quincy told her there was nothing wrong with what they put on Dennis's field. Jim

Moon, the Northwest regional manager, in Oregon, told her Cenex didn't have any responsibility for her horse problems. She called the corporate headquarters in Minneapolis.

I want to talk to the president of Cenex, Ruthann commanded.

What? the operator asked.

I want to talk to the president, the highest *guy up.*

She was transferred to the secretary for the president of Cenex/ Land O'Lakes.

Ruthann said, *I want to talk to* the *guy who's in charge of* all *chemicals and chemical plants for Cenex.*

She was transferred to another secretary, and finally she talked with somebody whose name she forgot, but who said he was in charge of the chemicals and knew about the Quincy Cenex operation.

Now look, you know you can't go dumping this stuff and then cutting it for hay! Ruthann told him.

Ruthann said she was told Cenex sprayed Curtail and Roundup on the field. She checked the label at the extension office, and neither of them were legal to spray on hay. If they were sprayed on hay, it'd be dead hay, so she didn't believe that.

She called the Cenex official back and he asked her what she wanted. Well, Ruthann wanted them to pay for her losses. She said she wanted about seven thousand dollars—the price of the hay, veterinary bills—and about five thousand dollars for the two colts that were aborted and the two well-bred mares that died and colts they could have produced.

A reasonable price. Especially since the one mare produced a colt that was sold for five thousand dollars. Sunshine, she probably had another eight or ten colts in her.

But the Cenex man didn't want to pay anything for any damages. He didn't want to admit to any fault. He offered to refund the thirteen hundred dollars price of the hay.

Ruthann just kept the hay. And she made sure to keep one of the older horses that ate a lot of it so the horse could be autopsied when it died.

OCTOBER 1991 — A FEW MILES EAST OF QUINCY

Tom Witte was a fifty-three-year-old farmer with two hundred acres and a hundred cows in a small dairy operation on Road I.5 between Quincy and Ephrata. He came from a long line of farmers. Tom's father had moved the family from North Dakota to the Columbia River Basin in 1956. It was a time when Grant County was spending $360 million—borrowed money, bonds fifteen times the county's net worth—on dams spanning the river. Grant County aspired to be the "Pittsburgh of the West."

The Wittes never got rich and they never even saw the real Pittsburgh, but they raised twelve children on the family farm. Tom took over in 1974, and he had been farming on his own since then.

Tom's primary crop was alfalfa to feed his cows and to sell to other dairies. He would rotate the alfalfa with wheat or corn every four or five years. Tom got through the Farm Aid years okay. In 1989 he took over the operation of a neighboring dairy and he was farming six hundred acres, mostly in hay.

But then, as he said, "I found out why all the previous operators of the dairy facility had failed, and I also discovered that operating a dairy facility is a whole lot of hard work."

Witte had a disastrous year in 1991. "A train wreck," he called it. The alfalfa was lush, but the red spring wheat, silage corn, and grain corn all yielded about a third of normal levels. He lost about three hundred thousand dollars and filed for bankruptcy.

"You always blame yourself, you know," Witte said later. "You always think you screwed up. But then it wasn't just the crops. Then I started having all these weird problems with the cows."

Six of his cows got sick and died. In the three that were tested, a veterinarian found cancer.

Witte had been buying his fertilizer from Cenex. The company put a thousand-gallon steel tank by Tom's irrigation spigot and filled it up as needed.

In 1991, Cenex was replacing the steel tanks on farms all over the county with smaller plastic tanks of concentrated liquid fertilizer to feed into irrigation water. As it happened, Witte's field man, suddenly stricken with muscular dystrophy, retired that year.

And in the confusion of the field man's illness and Witte's bankruptcy, nobody ever came out to pick up Witte's steel fertilizer tank. Years later, it would reveal a secret.

As 1991 drew to a close, Dennis DeYoung could be found walking around with one untucked shirttail, unshaven, and talking to himself.

The Cenex people said he was crazy, obsessed, nuts. People would duck into the restroom when they saw him coming.

Dennis had a grain mill sitting by his double-wide, but he didn't have any business for it.

He caught it from Marilyn for spending so much time on the subjects of waste and land.

Dennis was faced once again with a decision whether to take back his land from Cenex. He didn't want to take it back for the same reasons he had refused to turn the water on at the start of the year. He tried to get Cenex to rent the land again, but the answer was no.

Travelers wanted Cenex to buy Dennis's land and pay off the mortgage. "That may be the prudent move by Cenex," a lawyer for Travelers wrote. Otherwise, he warned, the mortgage company might have to sue Cenex for damages to the land. That would be something different: Travelers and DeYoung vs. Cenex in court. That would even the odds.

Cenex refused. The fertilizer company had the state on its side. "Cenex is advised that investigations by the State Department of Agriculture and by the Department of Ecology have failed to substantiate any of DeYoung's allegations."

After New Year's Day of 1992, then, the mortgage company

reached out to Dennis. He owed $46,524, and if he didn't pay immediately, they would file a suit to foreclose.

At Dennis's home, things were, if possible, getting even worse.

Dennis's brother, Dale, who had once run for mayor of Quincy, fell apart. Dale took off one day in his van, which he drove with a device for paraplegics, and disappeared. Dennis couldn't find him anywhere, day or night. The Grant County sheriff put out a bulletin and found him: Dale had driven to Spokane without telling anybody and he'd run his van up on the curb. He had sat there all day. Somebody finally called the Spokane police. They took Dale to the mental health unit for help. Dale would be under a doctor's care for many months to come. He told Dennis that he had a nervous breakdown because Dennis was talking to him too much about the toxins and the land. There was nobody else to talk to. After that, Dale and Dennis never talked again about the Cenex waste and the farm. But Dale's health kept sliding. Two years later, he would die at forty-eight.

Now Dennis felt doubly isolated.

A lot of other farmers wanted Dennis to be quiet. A group of Cenex growers gathered in the basement room at Paddy McGrew's restaurant in downtown Quincy one morning in December, Max Hammond presiding. Among other subjects they kicked around was Dennis DeYoung. Hammond told them Dennis was making false accusations and there was nothing to worry about. Nothing except for Dennis's mouth.

In the community of well-to-do farmers, Dennis's name was lower than dirt. He wished for the respect of the others, but he would not quit. He dug in deeper. Dennis tried to find other farmers who had chemical problems.

He heard about Willis Hendrickson. Hendrickson owned 427 acres south of town, used Cenex chemicals, and had crop problems. Dennis talked to Hendrickson. They were both wary. Then he started talking a little more. Dennis thought there was some corroboration. But then Hendrickson cut him off completely.

He looked at him like he was crazy. Dennis understood that look.

"When you're in a place like we were, five years of crops dying, somebody dumping a waste pit on your farm, you've lost everything and you say, 'Well, it's the fertilizer,' who's going to believe you?" Dennis would say. "Who in the *world* is going to believe you?"

CHAPTER 2

Suspicions

June 1992—Quincy, Washington

THE REALTOR ON THE TOWN COUNCIL and the city attorney whose word was taken as law were on the top of Patty Martin's suspect list. Her sixth sense told her not to trust them. She did not believe they said what they knew, and she speculated they had ulterior motives.

Why would they want to buy 1,595 acres by the golf course five miles out of town to pump and dump wastewater from Quincy's food-processing plants? There were better ways, better places, closer land, drier land, cheaper land. The irrigated land should not be used as wasteland. Patty couldn't have been more suspicious.

Five families owned land in the area between Roads Q and R NW and Roads 3 and 5. The realtor's company was buying options on their land for the Port of Quincy. Then Patty heard about somebody else who was talking with one of the very same families.

"I mean, none of it was making sense. Then somebody dropped Dennis DeYoung's name.

"I called Dennis, and Dennis started telling me about how his land was contaminated, and I told him about this acreage out there that the city was going for.

"And he said well yeah, he was comparing notes with Willis Hendrickson, and then the city put options on his land and Willis stopped talking."

Something clicked in Martin's mind.

It was, she believed, the secret reason they wanted to buy this piece of land more than any other.

Holy cow! That's what's going on! The land is contaminated!

She thought some company might have used the land by the golf course as a toxic-waste disposal site like Cenex on Dennis DeYoung's land, calling the waste fertilizer, now using a public agency to buy the land and flush it with millions of gallons of water.

She learned Willis Hendrickson had been paid $4,270 a year for two years to give the port district the option to buy his land for the handsome price of $640,500. No wonder Hendrickson stopped talking with Dennis.

The only way to buy silence is to buy the land! Patty said to Dennis. She smiled at him knowingly. He nodded.

Poor, lonely Dennis had finally found someone to talk to.

After that, Dennis and Patty talked almost every day.

Patty saw in Dennis a sad-eyed outcast with everything riding on one quest, and someone she might be able to help. Over time she learned he was a religious man, sometimes philosophical, intelligent but not book smart. He would misinterpret things he read, and she would tell him, "No, it doesn't mean *that*."

Dennis had this nervous energy. He grabbed his hands together by his belt buckle when he stood and talked, as if the only way to keep the hands under control was to hold them together. And he had the most distinctive laugh.

Patty would smile as Dennis cackled madly over various ironies, absurdities, and injustices he perceived. Dennis's high-pitched cackle reminded her of the sound of especially excited crows.

"You have to laugh or you'll cry," he'd finally say, wiping his blue eyes and shaking his head.

Dennis saw in Patty a figure both smart and caring. Patty had an athlete's energy, a college graduate's intellect, and a mother's sympathetic nature. She'd been back in town only five years and was already a community leader. She offered this all to Dennis DeYoung in an intense, platonic friendship like he'd never known before.

So these two people teamed up against the powers that were. To call them an odd couple would be an understatement. The Amazon woman who'd never say die until the final buzzer sounded, and the laughing, pear-shaped farmer whose eyes pleaded for understanding.

Dennis loved to theorize, philosophize, and complain about the waste pond's connection to his poor crops. Patty took what he said and made connections to other people and companies. She thought about her father's cancer again. Almost every day Dennis and Patty talked about chemicals and cancer.

Sometimes their spouses wondered if they could talk about anything else.

Together they developed a wider theory about possible illegal trade in hazardous wastes.

They decided the way to make sense of it all was to follow the money. Who was making money by using farmland to dispose of toxic wastes? They started to think a lot of people were in on the black market. They didn't know whom to trust, and they tried to be careful about what they said.

The theory embraced Dennis's misfortune and Patty's distrust.

Why did Cenex need such a big pond? The pond was by the railroad tracks with trains coming and going in the middle of the night. They thought Cenex had been unloading toxic wastes from railcars, mixing it in the pond, and shipping it out as fertilizer product. It made a certain conspiratorial sense. Nobody thought of fertilizer that way.

There was one big problem with it—no evidence at all.

And whose toxic wastes would they be?

It so happened the most poisoned waste site in the Western world was a one-hour drive south of Quincy, Washington. The Hanford Nuclear Reservation was a thousand times Love Canal, most comparable to the nuclear accident site at Chernobyl.

Hanford's operators had a long and shameful tradition of lying to the public. The 540-acre reservation was the nation's most tainted nuclear site. Hanford had been part of the Manhattan Project, refining plutonium for the atomic bomb. In the fifties, Hanford released clouds of radioactive gases under cover of night, exposing thousands of unwitting human guinea pigs, now known as Hanford Downwinders. Hundreds had filed claims for cancers and birth defects.

Hanford buried its worst liquid wastes in underground tanks, many of them leaking, the radioactive plume moving as silently and surely as gravity through underground water to the Columbia River. It was now the biggest cleanup program in the history of man. Along the way, Hanford lied about the radioactivity of wheat grown across the river from the plutonium plants and lied about the radioactivity in jackrabbits.

Once Patty called a citizens' group called the Hanford Education Action League to find out if waste left the site. *Every day, by rail and truck,* she was told. Patty asked them where it went, and they said they never thought to ask. But Patty and Dennis could never prove that Hanford nuclear wastes went to farmlands.

Their suspicion focused instead on a smaller company, International Titanium Incorporated, which happened to be located midway between Hanford and Quincy and which supplied cast-metal parts and exotic composite materials to Hanford, Boeing, and other customers. The company had enormous problems with its waste.

Dennis and Patty clipped articles from the newspaper. They set out to investigate.

Dennis knew somebody who said he'd seen workers bury fifty-five-gallon steel barrels from the titanium plant with a bulldozer. The plant was forty miles west of Quincy, near Moses Lake, less than a mile from the regional distribution center for Cenex/Land O'Lakes. By now Dennis knew cobalt and other metals associated with radioactive wastes had been found in the test on the Cenex rinse pond in downtown Quincy—the product spread on his farm.

Dennis filed a public-records request with state agencies. Poring over the paperwork, Dennis and Patty noticed International Titanium had admitted to improper handling of thousands of barrels of unlabeled hazardous wastes. State regulators gave the company an emergency permit to store the wastes with a September 1984 deadline to remove or recycle them.

Right away Dennis saw the date roughly matched the start of his own crop problems, and his father's, and his father-in-law's. Connected? He thought so, but had no proof.

Patty got waste generator reports from a state agency. They showed ITI had once disposed of sixty-three thousand gallons, and Dennis remembered the Cenex pond held about the same amount. Connected? No proof.

The company had a dozen employees working to dispose of four thousand drums of titanium tetrachloride, a liquid waste from the refining process. It sent some of them to the licensed hazardous-waste landfill in Arlington, Oregon. The landfill costs were astronomical.

Dennis carefully highlighted other parts of the report showing quantities he could hardly imagine: 7,552 barrels of solid wastes from the condenser process; 1,334 barrels from the filter system; 250 barrels from a chlorinating process laced with radioactive zirconium, uranium, and radium 226; 500 barrels of "miscellaneous" waste.

Dennis shook his head as he slowly read the reports. With his dyslexia, it took hours and hours.

He found letters in the file between a titanium manager and a department of ecology official named Jim Malm. By 1986 the company had 17,256 barrels of condenser waste. Colorado experts on recycling heavy metals advised International Titanium.[1] They wanted to recycle 5 barrels of waste a day, every day for two years, until they were all gone. By recycling instead of storing the wastes, the company would save two hundred dollars a day in the cost of barrels alone.

ITI also wanted to recycle 3,800 barrels of another waste by removing the hydrochloric acid and sending the radioactive residue to another company to extract vanadium.

Dennis wondered what they did with the acid. He knew there were a lot of acids in fertilizer.

From the correspondence, Dennis saw that state officials were torn between being tough and trying to help International Titanium. Its parent company, Wyman-Gordon Company, was already named in Superfund cleanups in Monterey Park, California; Cedartown, Georgia; Palmer, Massachusetts; Harvey, Illinois; Swartz Creek, Michigan; Leicester, Massachusetts; and Monroe, Michigan. Wyman-Gordon also had to pay $5 million to clean up its Worcester, Massachusetts, facility and $1.8 million for safety violations at a Texas forging facility.

On the other hand, Wyman-Gordon was a leading maker of titanium parts for airliners, fighter jets, and power plants worldwide, and one of its best customers, Boeing, was Washington's biggest employer.

In the end, the paperwork showed, state officials favored the helpful recycling approach. The titanium company was allowed to call a lot of its wastes "recyclable byproducts." In this way, ITI avoided millions in expenses for storage and treatment, and Malm avoided filling up landfills and driving off jobs.

This is what they do. They dump it on the farmer, Dennis said.

They're putting it all over the Columbia Basin, Patty said.

The enterprise changed when Wyman-Gordon announced plans in 1987 to close International Titanium after just three years of troubled operation. By dint of closing, the so-called by-products became simply wastes. The company could no longer spread them around without safety tests and record keeping.

By reading all the paperwork from state files, and talking with each other every day, Dennis and Patty learned how a hazardous material could be called a waste or a product—the same stuff—depending on what somebody planned to do with it.

Dennis and Patty had talked to anybody and everybody for two years, often with the word "help" in the conversation. Now they started trying to be more discreet. They started hearing about secret truck routes and night trains. This was huge. They thought their lives could be in danger. They had no idea where it all would stop.

Dennis, Patty said, *we have to control this imagination of ours.*

They kept studying and digging and asking questions, but they didn't share their suspicions about the big picture with many other people.

Patty shared with Anthony Gonzales. She trusted him. Tony worked with her husband, Glenn, at Lamb Weston, and was a high-school classmate of Patty's sister Mary. Tony encouraged Patty to keep asking questions about the plan for the industrial-waste site by the Quincy golf course.

Patty shared with Dwight Gottschalk, too. He was Glenn and Tony's boss as operations manager at Lamb Weston. Gottschalk was as tall as Patty and wore his silver hair trimmed short. He listened carefully and encouraged Patty to keep asking questions about the industrial-waste plans. The more we know, the better, he told her.

And Patty shared the suspicions with Glenn, of course.

"That could be," he would typically respond. "Maybe so."

One day Patty and Glenn drove down to the railroad tracks on the east side of town. It was less than a mile from their house. The tracks were lined by chemical and food-processing companies. They stopped on Division Street by a building marked "Cenex." Patty pointed to a plot of land between the street and the railroad track.

There was nothing there. The land was bare.

The waste pit, once the size of a big swimming pool, fifty-four feet long by thirty-six feet wide and five feet deep, had been filled in and covered with dirt. There was nothing to see.

But the topsoil was pale and fine and it stunk. Patty knew how the wind blew in the Basin. She could easily imagine dust clouds swirling with every chemical in the dirt.

They were standing in a light breeze two blocks away from Quincy High School, and across the street from the school athletic field.

Patty's eyes darted after every car that passed. She didn't want to let

Cenex know she was on to them. Patty was worried for her family's health. She saw growing evidence of a pervasive trade in hazardous waste endangering food.

But nothing she said made a difference to the people in Quincy. They just would not believe toxic waste was being used as fertilizer.

Glenn, too, was less inclined to suspect so much. Glenn and Patty looked a lot alike in their open-faced all-American personas, but Glenn was less stubbornly independent. He tried to get along. He wanted to rise in the business world. Patty learned how Lamb Weston collected the peels from potato processing and fed them to cattle. Glenn said that was good for the company and good for the cows.

Patty wondered about the pesticide concentrations in the peels. Patty wouldn't let her children eat the peels of baked potatoes. Glenn ate them peels and all.

Dennis and Patty kept an eye on companies they saw as cogs in the industrial-waste machine. One was a small company at the airport that sold liquid nitrogen. All it needed to be a fertilizer maker was to pay a twenty-five-dollar license fee every two years. Another was a giant lead and zinc smelter, Cominco, in British Columbia, six hours north of Quincy. Patty watched the railcars roll through town. She suspected they were moving toxic waste.

During this time, Cenex worked with state authorities to clean up the rinse pond site, and the state department of ecology named Jim Malm to supervise—the same Jim Malm who'd worked with International Titanium.

Too much coincidence? Patty and Dennis thought so.

But Malm started his job with an action Cenex opposed. He signed an order classifying the pond chemicals as "dangerous or extremely hazardous." That ended the semantics argument over whether the chemicals were beneficial nutrients. They were not. In April, Malm signed the final plan for the site, giving responsibility to Cenex to assess the toxic hazards, evaluate the health threat, if any, and clean it up.

Patty didn't trust Cenex, and now she didn't trust Jim Malm or the state department of ecology, either.

We need to get a sample off that site, Patty kept saying to Glenn.

One day he replied, *Patty, stop talking about it. Let's just do it.*

How? Patty asked.

Glenn cleaned the inside of a vacuum cleaner tube and drove down to the Cenex lot with Patty on a Sunday when nobody was around. He pushed the tube down hard into the ground several inches deep and pulled it up. Patty held a plastic bag open, Glenn tapped the dirt from the tube into the bag. The core sample reeked of chemicals.

Somebody drove up in a blue pickup truck. Glenn and Patty didn't recognize the driver. He asked Glenn what they were doing. *Just looking around,* Glenn said casually. *Any problem?* The man said Cenex had problems with kids rolling fertilizer tanks into the street. Patty thought that was a lie.

They took the dirt sample home and put it in the garage. They never had it tested, though, because Dennis took his own sample of dirt from the Cenex land a short time later and sent that to the lab. Patty was glad to throw out the bag of dirt. It had stunk up the garage and given her headaches.

Another school year ended. Patty cooked and cleaned and played with her children and taught swimming in the summer. And she read everything she could find on industrial wastes. One day she found a report from an engineering firm to the city describing details of the 1,595 acres proposed to receive wastewater from the food-processing companies. The report listed "the annual loading rate" of numerous toxic metals from the waste on the land.

Where is all that coming from? Patty said.

Don't you see? Dennis said. *They're going to push it through the soil with the irrigation water. Just like they did out at my place.*

The proposed wasteland was adjacent to the Quincy Golf Course. Patty knew the owner.

Lenore Annabelle wasn't getting rich with a golf course in the middle of nowhere. She harbored a dream to buy land around the course, put in manufactured homes, and create a retirement community in the desert like Scottsdale or Boca Raton. But Lenore's plan was opposed by some of the same people Patty and Dennis had suspected as "players" in waste disposal, including members of the town council. Patty thought there was a hidden motive.

It dawned on me when I talked to Dennis about them trying to buy the land next to the golf course," Patty said later. "They stopped it from being developed *because* they thought it wasn't safe to live on the land, or they knew it would have to be tested before it could be developed."

Patty poured out her suspicions to Lenore at the golf course. Lenore's son, then thirty and a golf course worker for nine years, had testicular cancer. One of the ingredients found in the rinse pond, cadmium, causes testicular cancer. There was no evidence the rinse pond material had been put anywhere near the golf course, but it made Patty suspicious.

Cadmium is a mobile and insidious poison,[2] efficiently pulled from the soil by roots but nontoxic to plant life, innocuous until it enters the food chain of animals. Cadmium can cause cancer and birth defects in people with trace exposure at just the wrong time in biological cycles. It can emerge years later in maladies of the kidney, liver, heart, testicles, pancreatic system, bone, and blood. Ninety percent of human exposure to cadmium comes from food.

Lenore didn't believe her son's cancer came from farm chemicals. "My goodness, when I was a child, I played in orchards with lead and arsenic. I don't think anything around here caused his cancer." And Lenore thought Patty was out of line scaring people about farm chemicals. "I feel sorry for the farmer having to battle everybody, the environmentalists, the water people, the health people."

But Patty talked to Lenore for hours. Before long, Lenore was taking a scoop with her when she drove her golf cart around the course, stopping every so often to take samples of dirt. She took them to Patty, and Patty sent them to a laboratory for testing.

Patty thought she picked up another clue during an unlikely opportunity. She was volunteering as a class helper for her daughter's fourth-grade field trip to the Quincy Lakes, part of a collection of geological formations referred to as "potholes" south of town. As the school bus drove past the golf course, Patty looked over at the 1,595 acres the city wanted to buy for industrial sewage, and at the green Quincy Lakes in the near distance. She'd always thought that would be a stupid place to put contaminated waste because the runoff would flow into the Quincy Lakes. But another parent on the field trip told her there was an unusual divide in the land gradient right at the place they were looking. One side, the runoff would go to the Quincy Lakes; the other side, where the wasteland ended, it would go the other way, toward a distant reservoir.

Something clicked in Patty's mind. The city was already sending wastewater effluent to the reservoir. Any contaminants added to that stream would, therefore, be almost undetectable. She thought it would be a perfect spot to wash pollution through the soil.

Everything about the golf course plan seemed absurd—to relocate families, buy their houses, use irrigated farmland as an industrial-waste site when there were better sites available—unless there was a hidden motive. After the field trip, Patty finally thought she understood. The pieces fell into place. She thought it was a cover-up of the industrial waste that was already put on the land.

She thought they would use the wastewater—more than four million gallons a day—to purge the land of contaminants. And she thought it was a setup for Lamb Weston and possibly Simplot, whose water it was, to take the blame if the reservoir died or the groundwater was contaminated.

Patty and Dennis worked it out.

One night, Patty told Glenn.

Could be, he said. *Maybe so.*

Carmen Weber, a town leader and former classmate of Patty's, agreed the city wasn't doing enough for Lamb Weston. Carmen asked Patty to think about running for the town council. Patty was always attracted to public service. She was an inside player and a member of the starting team in basketball, and she wanted to be that way in the community.

The council had a one-year vacancy. Patty was one of nine people who applied. Glenn was dubious, thinking she ought to just spend her time at home with the kids, but Patty won him over with the argument that she would bring in one hundred and fifty dollars a month for attending council meetings that she was already attending for free.

But she lost. In a vote by the other council members, Dick Zimbelman won the job. He was a contractor, farmer, and former fire chief. Patty liked Zimbelman. He was okay. He got it. She could talk with Zimbelman and another council member, Jim Hemberry, about her suspicions, and they encouraged her to keep digging.

Patty convinced herself it was a relief not to have been picked. Just as well to be on the outside.

Stubborn Patty, more and more intrigued with the questions no one could answer, worked even harder. There was so much to do: Dennis to help, the Cenex dust, the golf course scheme, ITI toxics and Hanford in the distance, and the effect of all this on fertilizer and land and food.

Glenn thought the telephone was glued to Patty's ear. She couldn't deny it. "I'm on the phone all the time because of this," she said. "My kids are growing up around me."

Eventually Patty's questions about contamination on the land by the golf course seemed to have an effect. The council voted, three to two, against the plan to buy the 1,595 acres. Patty thought the town leaders had stopped pushing so hard when they found their secret leaking.

Patty played pickup basketball at the high-school gymnasium on Sundays. She always said: Big women who played power basketball had to hustle more than anybody else. She ran the full distance from end line to end line. And she liked to shoot. Sometimes Glenn and their son David played. Some days she was sore from being banged by the men and the teenage athletes, but she had fun and loved to compete.

Dennis and Patty talked incessantly, but rarely appeared in public together. Patty never wanted to talk with Dennis around big groups of people. Once he backed off between two big bushes while they were talking in the front yard. Paranoid Patty was getting Dennis worried about who was watching them.

The outsider role fit them both. Huddled like birds on a branch in a storm, they talked about everything they had learned, guessed, or imagined about waste and fertilizer in the Columbia River Basin.

Dennis had simply walked away from his own farm. He pruned the orchard but did nothing on the circle. He left the water off. All that grew were weeds. His only chance of holding Cenex accountable was to walk away.

Dennis drove a Mack truck. He hauled crops and grain all over the state. He got a job selling sweet-corn waste for J.R. Simplot Company. Off duty, he threw himself into investigating the local fertilizer companies.

Dennis's father, Jake DeYoung, also filed a complaint with the state in 1992. His barley had died and then the beans he had planted over that did poorly, and this year his three fields of corn plants turned purple. Jake DeYoung, like his son, blamed Cenex, though without proof. If chemical companies were going to dump on anybody, Dennis and Patty figured it made sense for them to dump on the poor and the DeYoungs. Nobody would believe them anyway.

"Right away we could see if you had this knowledge and somebody else didn't, then you had the power to choose who was going to be successful and who was going to fail," Patty recalled.

An inspector for the state department of agriculture was the first to take them seriously.

Cal Briggs listened intently while Dennis told him why he thought his crops were so poor in the late 1980s—blaming Cenex, the rinse pond material, the so-called fertilizer. It was more than a one-time deal, Dennis told Briggs. Cenex had been dumping waste on him and others for years, he said. And Briggs seemed to agree.

One day Dennis brought the state inspector to meet with Patty at the Martins' house. Patty thought Briggs equated her air-conditioned home and her boat in the backyard with middle class, money, and credibility.

Briggs took more interest in Dennis's complaint after that meeting because it was no longer just one broke farmer standing alone. Patty sat right by Dennis and described her own fears about industrial wastes posing as fertilizer. She had no financial self-interest in the complaint.

Briggs told her that might be happening, might be true. He said he'd found out the state department of agriculture had licensed industry waste as fertilizer, and he was disgusted they would do that. Briggs thought there were other farms that had been harmed. He told them he was looking for somebody to help him take this on.

Briggs came to Quincy several times in May, June, and July of 1992. He rode around with Dennis and took pictures. Much of Dennis's one-hundred-acre circle was still dead, two years after the company had spread material from the rinse pond there. The field wasn't being farmed or watered, but Briggs thought that wasn't the only reason for the poor condition.

Briggs took careful, detailed notes to describe the misshapen, stunted weeds he did see: China lettuce, kochia, Russian thistle, lamb's-quarter, and buckwheat. He thought it looked like chemical damage. The soil pH had dropped in two years from 7.9 to 5.6. The EPA test result on the waste pond showed a lot of other pesticides they hadn't known about earlier.

But then, after a while, Briggs disappointed Patty and Dennis by telling them he wouldn't investigate further. He wished them luck, but Briggs said he couldn't rock the boat because he was too close to retirement.

The inspector wrote at the end of his official report, "There was not any conclusive evidence found that Cenex routinely applied contents of the pond to farmland in the Quincy area."

Dennis DeYoung drove up to the Quincy Cemetery with his friend Russ Sligar to pick out a burial plot. Russ's wife, Terry, had died of cancer. She was a mother of four. She was thirty-seven years old.

Later, Dennis helped Russ take samples of his pea and potato crops to test for chemicals. Terry Sligar had worked in those fields. She had changed sprinklers in bare feet.

The test results showed high levels of arsenic, barium, lead, titanium, and chlorides.

Dennis wanted Russ to join in his fight against Cenex and toxics in fertilizer and to help stir up the EPA.

But Russ was sick and brokenhearted and didn't want to fight.

Over time, many of the people in whom Patty had confided turned against her.

Her suspiciousness about industrial waste cast her as an extremist. Sometimes people just walked away from her when she started talking about it. Her husband's boss, her former confidante Gottschalk, thought Patty was more and more paranoid. It was an opinion shared sometimes by Glenn as well, though he usually stood up for her right to express opinions or, more often, stood silent, not wanting to take sides in an argument.

At home, when Patty said it wasn't fair, Glenn replied, "Who said life's fair?"

When Patty talked incessantly about hazardous waste in fertilizer, Glenn replied, "What makes you think that anybody cares?"

Patty had some good reasons to be paranoid. A local phone company manager actually did monitor some phone conversations in Quincy.

You can't tell anybody, but you know, this is part of his job, the man's wife told Patty and another friend, Kathy Knodell, the prosecutor's wife.

It was not part of his job to monitor the same phone line over and over, but the manager did it anyway. He told Patty about some other conversations he'd overheard, so it was fair enough for Patty to assume he might listen in on her telephone line, too.

Word got around. Secrets were not kept.

Patty refused to talk about some things on the phone, and during sensitive phone calls she would try to scare away the imagined eavesdropper by saying hello to him, though she got no response.

The lobby of the Grant County Courthouse was nearly empty, the sunlight blanched on linoleum floor, as two clerks walked by, reaching for their midmorning cigarettes. The sheriff, William Wiester, stood at the side of the lobby drinking coffee.

A man in a suit walked in and approached Sheriff Wiester. He greeted him like a friend, for they had worked together before. He was a lawyer working for the mortgage company Travelers, which held a $124,000 judgment against Dennis and Marilyn DeYoung.

The lawyer was the only person to show up for the final act in the foreclosure on the DeYoung farm. It was ten o'clock in the morning of the first Wednesday of September, the appointed time for the sheriff's sale.

There was only one bid. The lawyer said, *one hundred twenty thousand dollars*. Sheriff Wiester looked around the lobby one more time.

Sold.

The lawyer and the sheriff signed a form. Shortly, Travelers appointed a receiver to take care of the land. The receiver hired Cenex to farm it.

Of course, farming wasn't Cenex's business. Cenex sold chemicals to farmers. Why would it compete against its own customers?

In this case, Cenex made an exception. The company hired a crew to plant fall grain on the DeYoung circle.

Under the state foreclosure law, Dennis would have one more chance to get his land back if he could make the payment within a year.

Cenex wrote the cleanup plan for the rinse pond site. The report said, "No pesticide concentrates were knowingly dumped into the rinsate pond." The word "knowingly" was a lie. Cenex employees had admitted to dumping pesticides in the pond at Williams's orders.

Dennis wrote on a yellow sticky note attached to his copy of the report, "Who could dump pesticides and not know it?"

He thought Cenex was lying about the money, too. The company listed the $46,524 it paid to rent Dennis's land in 1991 as a loan instead of rent.

Dennis thought it was fraud. He wanted to file another complaint, and he didn't think the state officials were any good. He recruited the widower Russ Sligar to come along.

Let's go to the FBI, Dennis said.

Russ watched while Dennis put a couple pounds of his paperwork in a big old cardboard box at the grain mill. There were Cenex bills and state reports and letters and deeds and appraisals and articles and pictures and legal pleadings. Dennis carried it all to his pickup truck, and they went out looking for the Federal Bureau of Investigation.

First they drove south to Yakima, but there was no FBI office in Yakima. Then they checked the phonebook and drove east to Pasco, found the federal building, and walked inside. Dennis told a receptionist they wanted to talk with the FBI. She picked up the phone. An FBI agent came out of somewhere in back of the building.

Dennis poured out the story, showed him the box of paperwork. The agent didn't know what to do, so he called the field office in Seattle and introduced Dennis to another FBI agent, a white-collar crime specialist who set up an appointment a couple days later.

I think they sort of believed me, Dennis said to Russ on the drive home, steering west toward the sunset. *If you're totally crazy, I think those guys could recognize that.*

Soon, then, Dennis went over to Seattle with every document he had. He met the FBI agent there and talked to her about financial fraud and toxic waste. She referred him to somebody else, an agent with the Criminal Investigation Division of the Environmental Protection Agency in Seattle.

Dennis was getting the runaround.

Then Patty Martin phoned the EPA.

Or was it Patty Murray, the United States senator?

———

Some people confused the two. Martin was a foot taller than Murray, but you couldn't see that on the phone. One day in the spring of 1993, Patty called the EPA in Seattle. She said she was Patty Martin, but the exquisitely political EPA bureaucrat seemed to misunderstand.

By the tone of voice, the demanding nature, perhaps, and the similarity in names, the bureaucrat apparently thought he was talking to Patty Murray, U.S. senator from Washington State, who'd won election as a "mom in tennis shoes" concerned with health and the environment.

Patty Martin, nobody from Quincy, found the federal agency unusually attentive to her phone call.

And it was not many days later, on a bright Tuesday morning, May 11, 1993, when the feds descended on Quincy. They surprised everybody.

A dozen inspectors from the EPA Criminal Investigation Division, Superfund Response and Investigations Branch, and Technical Assistance Team, driving government-issue vehicles, pulled into town unannounced.

They were investigating at the request of Senator Patty Murray, or so they believed. Armed with search warrants, they drove straight to the Cenex site by the railroad tracks.

An agent asked to speak with John Williams. Under auspices of the Toxic Substance Control Act, he said, the EPA would be taking soil samples from the Cenex property. Williams said okay. Then he went inside his office and phoned the Cenex lawyer, Michael Tabler, and the regional credit manager, Nick White, and word got back to headquarters in Minneapolis quickly, blaming the incessant whiner, the bad farmer, the loser DeYoung.

"I don't know how Dennis pulled this off," the lawyer said later, "but he got Patty Murray to come to Quincy, and he got Patty Murray somehow to come to his corner. And the next time I know is ten or twelve people from EPA come to our site and with search warrants and hauling my people off to be interviewed like criminals."

They took files. They used a backhoe and shovels to dig a trench thirty-six feet long and two feet deep in the old rinse pond site. Three or four of the workers put on full-protection bodysuits that made them look like spacemen as they scooped up samples of the soil.

Dennis DeYoung was one of the first to hear. He drove over and sat in his car down the road, watching. It was the first time he'd felt joy in a long time.

At the time, Dennis wasn't sure which Patty had brought the Federals to town. He'd sent letters to Senator Patty Murray. "It was cool. They really sent the soldiers in here," he said later. Dennis called Patty Martin as soon as he got home.

Guess what? he said.

Martin drove down to the plant immediately to watch the action. She didn't stay long. She didn't know if she'd be safe from the chemicals or the Cenex managers.

Almost nobody else in Quincy knew the feds were visiting their fair town, or so Patty thought.

That afternoon, she stopped by Call Drug Store to fill a prescription and saw Dave Manning, the owner, at the back of the store. Patty was feeling vindicated, validated by the EPA. She walked up to Dave grinning like a little kid with a big secret.

I know something you don't know, she teased.

What's that? Standing on the pharmacist platform behind the counter, he looked at Patty eye-to-eye.

The EPA's in town pulling samples at the Cenex site, and they're wearing space suits, the full equipment!

Oh, I know, Manning replied.

The matter-of-fact tone caught Patty by surprise, and she felt deflated.

Manning wasn't even curious. A federal invasion of Quincy's industrial area, and he didn't want to talk about it. Patty wondered if the handful of people she and Dennis had started to call "the enemy" were ahead of her every step of the way. Patty didn't feel so tall anymore.

The EPA left Quincy in midafternoon to make the two-hour drive back to Seattle. The test results came back within a week. Three of the metals in the rinse pond sludge were higher than the cleanup level set by law.

SUMMARY OF ANALYTICAL RESULTS

Rinsate Pond Sludge/Soil Sample #93200179
Quincy, Washington
May 11, 1993
ppm (mg/kg)

ANALYTE	TEST LEVEL	TOXICS CLEANUP LEVEL
Beryllium	1.39	0.223
Cadmium	25.2	2
Chromium	360	100

Patty asked, *Where's the beryllium from?* She would ask over and over during the coming years and never get an answer. Beryllium, she thought, was the key to this Pandora's box.

Dennis wondered, *Where's the chromium from?* It had no place in fertilizer.

The EPA also found 712 parts per million of titanium in the buried rinse pond. The EPA had not set a cleanup level for titanium, but it confirmed more of Patty's and Dennis's suspicions.

By coincidence, the next day a private laboratory in Idaho finished some work for Dennis DeYoung. The lab measured twenty-six chemicals in the old Cenex rinse pond and three other samples Dennis had sent in.

The titanium stood out again to Patty and Dennis. The lab found 1,560 parts per million titanium in an unattended Cenex fertilizer tank; 1,580 in dirt behind the Cenex store; 1,650 in dirt from Dennis's father's farm; and 2,640 on the empty lot where the rinse pond used to be. Dennis's sample contained four times more titanium than the EPA had found in another spot by the pond.

Both Dennis's and the EPA's measurements were hundreds of times higher than the highest level of titanium found in uncontaminated soil.

May 1993 — Quincy, Washington

Carmen Weber kept talking to Patty about running for office.

I'm not running. I'm not running. I have babies at home, a two-year-old and a four-year-old. I'm not running, Patty said.

Well, I'm going to call you every month till you change your mind. And I don't want you just running for the council. I want you running for the big one.

No, no, no.

Patty kept telling herself she would not run for office. She'd been willing to toss in her name for a one-year vacancy on the council. But that would be a four-year commitment. She was too busy: as a reading tutor, swim instructor, Cub Scout den mother, and soccer mom. She cooked and cleaned for her family. She had four children. Her husband didn't want her to start a political campaign.

No, no, no.

But another voice said, *Go for it.* The earlier loss had been a disappointment. That had been a council vote, inside politics. It was tempting to test her appeal with the larger group of voters in Quincy.

Carmen kept appealing to Patty's civic duty. The town needed her. The guys on the council, Dick Zimbelman, Jim Hemberry, they said they needed her. And by the last day of the week in which people could file for public office, not a single person had filed.

Patty went to lunch with Glenn that Friday. She didn't tell her husband she was running for office. Once again she asked his permission.

You're not running, he said, afraid she'd go overboard.

Patty tried to explain her thinking. *I could stay at home and be cynical about my city government, but cynicism breeds apathy, and one thing I am not is an apathetic person. That would kill me. And the other thing I could do is at least try to make a difference.*

Glenn listened carefully. The job was only part-time. She'd make a little money. She'd be happier.

What do you think if I just go over and if they need me, I run? Patty asked.

Okay, Glenn finally said.

He was, after all, a practical man. The only time Glenn stood in his wife's way was when he tried to check her during pickup basketball games, and then, often as not, she'd just shoot over him.

So Patty drove twenty miles to the county courthouse in Ephrata, her mind in overdrive—*maybe, maybe not*. When she got out of the car by the courthouse, she turned toward city attorney Lemargie's office, across the street, and waved, imagining Lemargie looking out the window and hoping she wouldn't run.

At ten minutes to five o'clock, then, Patty stood on the courthouse steps with a group of about six people debating who would run for what office.

Well, I'm here if you need me, she said to Dick Zimbelman.

We do, he said.

Mayor Debra Adams had walked by a few minutes earlier and told Zimbelman she hadn't decided whether she was going to run for reelection as mayor or for a council seat.

Zimbelman turned to Patty.

You have to run for mayor.

For mayor?

Yeah. It's you or Nick Todd.

Who's Nick Todd?

He was the man standing next to Zimbelman. Patty had never heard of Nick Todd. She thought he couldn't win. Patty, in contrast, was known at Q-CARE, the schools, the pool; everybody knew her. Maybe she was the only one who could beat Debra Adams.

Oh my goodness, Patty said.

Patty and all the others walked up to the clerk's office. The clerk locked the door behind them. Adams was at the counter filling out a candidate's form, shielding it with her left arm like a smart schoolgirl hiding her test paper.

As soon as Adams was done, Patty asked the clerk what Adams had run for. *Mayor,* the clerk said.

Patty looked at Zimbelman.

You gotta do it, Patty, he said.

I can't.

You have to.

I have four kids at home, Patty protested.

But she was competitive. She liked to shoot, not pass. Her team needed to send a player into the game who could win the election. She decided to take Debra Adams on one-on-one.

Patty shook her head as she filed the form. *Oh my goodness,* she said again. She'd never talked with Glenn about running for *mayor.*

Afterward, Zimbelman laughed. *I'm sure glad you did that, Patty, because if it hadn't been you, then I was going to do it.*

Zimbelman filed for a council seat. So did Tony Gonzales. Both were Patty Martin people.

At least for now.

As Patty drove home she asked herself, *What am I doing?*

By August of 1993 the DeYoung farm had a new owner: Cenex/Land O'Lakes. The local managers and the board of directors in Minneapolis had decided to redeem the $130,000 promissory note and take the land from Travelers.

"They run a farmer out of business, then they get his land. Now isn't that something," Dennis cackled.

Ruthann Keith noticed the neighboring land was once again lush and the irrigation no longer overflowed to create a stream down the hill. It was back to normal. Decent crops were starting to grow again on the hundred-acre circle. The poisons had been washed out of the root zone by Roosevelt's irrigation and the soil's natural healing powers.

Ruthann had given up the idea of getting money from Cenex for her dead horses. The leftover hay, bleached by sun and washed by rain, was no good as evidence.

"I kept the rest of that hay, and it's just sitting out there kind of melting down," Ruthann said. "I don't know what to do with it. I don't want to burn it 'cause then it'll blow all over and whatever's in it will blow all over, too. Maybe I should ask Cenex for disposal costs on the hay. But, ah, anyway. . . ."

Ruthann was busy and hardly ever saw Dennis. She figured if Dennis won his lawsuit then she could think about joining him again, but till then, she had plenty of Appaloosas to keep her occupied.

In early September, Dennis had his last chance to get the farm back. It was the end of the one-year period in which, by state law, a debtor who loses his property in a sheriff's sale can redeem it for the price of the debt. He could have bought his farm back for just over the $120,000 Travelers paid. But he still would have owed Cenex $130,000, and he would have assumed the responsibility for the tainted soil.

He let it go.

He hoped a judge and jury would help him.

———

The campaign for Quincy mayor was notable for the way euphemism and understatement masked the real issues the politicians talked about. Patty was no exception. Her brochure, absent a single word on industrial waste or contaminated land, was printed on pink paper:

IT IS TIME FOR A CHANGE

* * * * * * * *

VOTE
PATTY NAIGLE MARTIN
MAYOR FOR QUINCY

THE CHOICE OF THE
COMMUNITY

The brochure listed Patty's community service and spoke of street repairs and police hiring. And on the back page, the endorsements:

Anthony R. Gonzales, City Council Candidate, Unopposed, Position #4: "I anticipate a very productive, rewarding four years on the council with Patty Martin as the Mayor."

Jim Hemberry, City Councilman, Position #1: "Patty Naigle Martin will bring a fresh, creative, hardworking presence to the office of mayor."

Dick Zimbelman, City Councilman and Candidate Position #2: "Patty's honest, open approach to leadership will be a refreshing change on the city council."

Patty Martin was an engaging candidate. She went door-to-door, sometimes carrying two-year-old Eric. She was fresh, energetic, and loved to campaign.

Patty noticed a lot of children with mongoloid features while working in the schools and campaigning. She talked about it with a third-grade teacher. They thought it was caused by maternal alcohol or drug exposure during pregnancy. Patty doubted it. She thought those particular mothers didn't have enough money to buy much drugs or alcohol. She thought the mongoloid appearance may have been caused by something toxic in the diet or the dirt or the air of the farming town.

"It sure changes them from being victims of their mother's choice to victims of circumstance," Patty said.

But in public, at least, not a word about Cenex or the rinse pond or fertilizer or toxic waste passed her lips.

The people voted Nov. 2, 1993.

FINAL VOTE

QUINCY MAYOR

| DEBRA ADAMS | 330 |
| PATRICIA ANNE MARTIN | 649 |

The mayor's office was a six-by-ten-foot room with a window to the hallway and a green glue-down carpet.

It wasn't as nice an office as Patty's laundry room at home, but it held considerably more authority.

———

Dennis had lawyer troubles. His second lawyer had pulled out of the suit against Cenex, and Dennis was looking to hire a third just before the judge was to hear the company's motion to throw the case out on summary judgment.

C. E. "Monty" Hormel was an old country lawyer in Ephrata, well connected to a former state senator for the area, another lawyer who'd lasted twenty-five years in public office despite being a Democrat in redneck country. Hormel didn't care that Dennis was an outsider. He was intrigued by the case. At his customary rate of one hundred dollars an hour, he figured it would cost five thousand to ten thousand dollars to fight the motion for summary judgment.

Dennis said, *Go for it.*

The first thing Hormel did was file a motion of discovery. Cenex had to search for all the paperwork on the old rinse pond. Nick White got the duty. He'd been promoted from credit manager to area manager and would soon be named regional manager over five hundred employees.

White found the incriminating evidence.

He thought about throwing it away. He was holding the only copy in his hand. Nobody had seen it but him and two other company managers. It could have disappeared. For a while, White considered losing the document, then he put it in an envelope and sent it to the lawyers.

It was the first document that would not only incriminate Cenex, but shine a light on the industry practice of saving money by calling hazardous waste by another name: fertilizer. It was the memo saying Cenex could save $170,000 in hazardous-waste disposal costs by putting the rinse pond material on the DeYoung farm. They could just call it a product and spread it on the land.

Unfortunately for Dennis, the document was found too late to enter in his filing in court.

Cenex moved for summary judgment. Michael Cooper, a visiting judge from Kittitas County, read the sworn statement from John Williams saying Dennis had approached him about putting the pond contents on his farm. A statement from Larry Schaapman said it was Williams's idea, but Williams denied that. Judge Cooper read the sworn statement from Dr. Max Hammond saying the waste pit hadn't hurt Dennis's farm. He blamed cool weather and poor farming.

Dennis and Monty Hormel responded with their best evidence at the time: reports by the university professor and the state department of agriculture describing the chemical damage to his land, and test results from the EPA showing beryllium, cadmium, chromium, and titanium in the Cenex waste pond. The EPA report showed Cenex was the target of an ongoing criminal investigation by the EPA. But Dennis hadn't received the incriminating document from Cenex in time to give it to the judge.

Judge Cooper tossed out Dennis's lawsuit in May 1994. The judge ruled that Dennis hadn't proved damage to the land beyond 1991, when he could have taken it back and turned on the water and paid the mortgage. Therefore, Cooper ruled, there were no real questions of fact for a jury to decide. Cooper signed a decision that supported all the Cenex claims.

That wasn't all. Cooper reproached Dennis for not paying the $130,000 promissory note for Cenex farm chemicals in the 1980s. The judge issued a summary judgment for Cenex on the promissory note.

Cenex could not have received any better ruling.

Dennis was devastated. He could hardly look up from the ground as he walked around his shop the next day.

Hormel was surprised at the one-sidedness, the totality of defeat. He immediately started work on an appeal. He told Dennis to get ready for a long wait, two years at least, and he warned him the appeals court almost never overturned a summary judgment.

So now Cenex owned Dennis's land, and Dennis owed Cenex $130,000 plus $79,000 interest.

Patty vowed to fight harder. She had not quit. Dennis became a tragic figure in her eyes. He was philosophical in defeat. It made Patty more determined. "It's just so wrong!" she said. The Martins had so much and the DeYoungs had nothing. Dennis was her best friend outside the family.

And he needed her more than ever. Sometimes just seeing her gave him heart. Patty always tried to cheer him up. She told cornball jokes. He practiced gallows humor.

"You've got to laugh or you'll cry," Dennis said, obviously near tears. It was August, 104 degrees in the shade, and he dripped sweat working in the grain mill and driving the old Mack truck.

Patty stopped using her dining-room table for dining. She covered it with papers, reports, articles, legal filings, and laboratory test results.

She studied toxic waste. The dining room was Patty's war room. The family telephone bill, previously $30 a month, except when she'd been trying to find a cure for her father's cancer, was now running $100, $150. Patty reported every success and failure to Glenn and Dennis. She was overflowing with energy and resolve.

Glenn told her to stop running around the house. *Walk.*

In her wildest dreams she couldn't have come up with a more nefarious plot than disposing of titanium, beryllium, or radioactive wastes through fertilizer. But it was true.

Patty was good at recalling facts and figures, names of chemicals and maladies, and she spoke in a clear, confident voice with Middle American tones. She kept coming back to biology, medicine, health, children. Patty used a shotgun approach firing facts and suspicions at anybody who would listen; anybody, however briefly, she thought could help.

After harvest, work slowed for Dennis and he devoted more time to the cause. He spent hours going through old newspapers, court files, and government reports.

It was no wonder Marilyn thought about leaving Dennis. She wanted a nice house and a new car. She wanted a regular husband, not bankruptcy, bitterness, and the depressed, hopelessly obsessed man she had.

One day Dennis told Patty that Cenex might forgive his debts if he would drop the case pending at the court of appeals. The appeal was a long shot anyhow.

Patty was insistent.

Dennis, whatever you do, you can't *settle out of court. Because you don't know how many other people have done that over time, you know, had their silence bought. And people have a right to know.*

Larry Schaapman had a beautiful voice. The handsome baritone sang out in the choir of the Quincy Free Methodist Church. He thrilled the congregation with a solo from the dais. He raised his hands to Heaven and praised the Lord.

Dennis sat in the pew, glaring. It was so obvious, people started talking about Dennis's glare. A visiting pastor asked, *Why is this man looking at that other man like that?*

Marilyn told Dennis, *Don't look at Larry.*

How can I not? Dennis replied.

Larry told people the judge had tossed out the lawsuit but Dennis still wouldn't drop it. He said Dennis was taking it too far.

Sometimes Dennis hated Schaapman, really hated him, for renting his circle of land without telling him about the side deal with Cenex to dump the so-called fertilizer product. And he hated Larry for making twenty-nine thousand dollars risk-free from Cenex. For blaming Dennis for the poor crops. For telling people Dennis was a liar. For being successful while he was broke.

Marilyn tried to tell Dennis they were going to church for friends, family, and Jesus, not for Larry Schaapman. She thought God would be the one Larry would have to deal with. "It was hard to watch that," she said later. "Larry being in leadership and standing up in front of the congregation singing like his life was just so great, and here we were. I felt like it was all a lie."

Dennis told Marilyn he wanted to let go of his hate because he knew it was a poor way to live, but he could not let go. He wanted to expose Larry and restore his family's standing among the church members.

Dennis talked frequently with Pastor Ed Burns. He insisted the court judgment was wrong and he, not Larry, was the victim of injustice. He invited the pastor out to his house to see the evidence. Once Burns came out and Dennis showed him the statement by Larry Schaapman in which Schaapman had agreed the product in the pit was a waste and Dennis's land was used for disposal.

See? He knew, Dennis said.

Pastor Burns said that didn't make one bit of difference. There was no proof Larry had known he had done anything wrong. Dennis hurt himself. The pastor told Dennis his fellow parishioners would like him to stop talking about the chemicals, the circle, Cenex, and Larry Schaapman.

If he had truly been wronged, the pastor said, *Forgive.*

Too soon, Dennis replied. *First confession, then forgiveness.*

⬤➤

The pastor's closest advisers gathered in a house next door to the church to listen to Dennis and Larry talk on a cold afternoon the week before Thanksgiving.

Apprehensively Dennis and Marilyn walked up to the back door. Marilyn's shoulders slumped. She had so many woes. She was taking care of her dying mother, whose cancer returned a year after a mastectomy.

It's a hard time to go, Marilyn told Dennis, *but I really want to talk to them and be able to talk to Larry's wife, maybe say, you know, "Let's get this straightened out."*

The pastor's cabinet sat on cheap, padded chairs around an oblong, Formica-top table in the main room of the house. Merle Royer, owner of the Coast to Coast store, was there; Marilyn had been his neighbor growing up. Ron Huxtable, a state patrol trooper. Brad Bierlink, a neighbor of the DeYoungs'. Ed Moore, whose daughter had died of a brain tumor. Marilyn knew some of these people had been through a lot and would somehow understand. Larry and Julie Ann Schaapman were already there when the DeYoungs walked in and sat down.

Everybody in the room knew that Dennis had sued Larry, Cenex, and John Williams, and that Dennis had lost the case on a summary judgment. They knew Cenex held a two-hundred-thousand-dollar judgment against Dennis for the bills he owed. They knew Dennis had filed a snowball's-chance appeal with the state court of appeals. They knew summary judgments almost never get overturned on appeal.

So Larry was feeling pretty secure. Dennis spoke first. He spoke quietly as he pleaded for understanding.

He'd suffered. His family had suffered. All he wanted was an admission by Larry that he'd done him wrong.

Dennis was surprised by what Larry said: He talked about the benefits of using leftover fertilizer.

Brad Bierlink said it made sense to him to use the contents of the Cenex rinse pond. He complimented Larry on making good use of left-over fertilizers.

The man got paid to dump a waste bin on me, Dennis thought. *It's like complimenting Jim Jones on the wonderful punch and forgetting about the cyanide.* "Good punch, Jim! You just made a nutritious drink!"

Larry said John Williams and Cenex had done nothing wrong, either. He said the weather was too cold, he hadn't farmed the land well enough, and Dennis was partly to blame because the irrigating circle had fallen down.

Not my fault, Dennis said.

It was your responsibility, because you fixed it, Larry said.

Well, I can't win, Dennis said.

The closest Larry ever came to apologizing was to say, *If I have wronged you in any way, I am sorry.*

At that, Pastor Burns turned to Dennis and Marilyn.

Now, is this okay with you two? the pastor asked.

Marilyn broke down and cried. After a while she composed herself enough to speak.

The thing that really gets me is people won't come up and ask us what's really going on here, she said. *It's like they don't want to know, and it's so weird because you feel like a family here, but people aren't coming up to ask what's going on. So we feel like we're the outsiders.*

Marilyn thought about her husband's struggle for respect. She thought about her mother, dying of cancer. She thought about how sad she felt now all the time. She needed friends. She needed the church's support.

You guys, here we are supposed to be this Christian family of people that care for one another, and we are just cast aside.

Other people cried, and apologized for not being more supportive. They said it was a hard situation. They didn't know who to believe or what side to take.

We're sorry.

We'll pray for you.

Burns turned to Dennis and asked him if he could forgive Larry.

I can't, Dennis said.

There was a deep, sorrowful moment, which struck the DeYoungs as sincere, and the meeting ended. Several people, including Marilyn, cried quietly. Dennis and Marilyn walked outside, climbed up into the old Ford, and drove away from the church.

"What *right* do they have to pray to something that they're not willing to do?" Dennis said later. "What right do they have to ask God to do something that they can do, but refuse to? I told them we have paperwork that shows this. They can look at the paperwork. It was just so amazing, 'Well, we really don't have time to look.' Oh, you can ask God to look and take care of this, but you don't have time."

"Very strange," Marilyn said.

"They were trying to appease us, pacify us, instead of the search for the truth. That wasn't even on the agenda," Dennis said.

Marilyn wasn't sure what to think about Larry Schaapman. "I don't know if he feels like he did anything wrong, in his own mind," she said. "I can't believe that someone would do something like that and not eventually say, 'I'm sorry. I did wrong.' "

They could not square their anger at the pastor's cabinet with their views that most of the church members were good people with Christian values. Dennis and Marilyn decided the pastor's cabinet took sides for Larry and against Dennis because Larry was more active in the church. Dennis and Marilyn were always busy, working to pay the bills, while the Schaapmans had more time to get involved.

A week later, they found out more about that. The DeYoungs took their children to the 9:45 A.M. Sunday school at the Free Methodist church, and when they met up before the 11:00 A.M. worship service, one of the kids had a brochure on upcoming classes. There was a new class on finances. The pastor appointed the teacher. It was Larry Schaapman.

What!? Dennis exploded.

Marilyn read it and said, *You've got to be kidding!*

They could see someone teaching a church class on finances—and who knows, maybe they'd even attend—but of all the people. . . . Larry could have had the decency to let the problems get sorted out first, they thought. Marilyn said it was very disrespectful.

That's it! she said.

So the DeYoungs stopped attending the Quincy Free Methodist Church.

Marilyn missed the church intensely. Some of her friends asked her to please come back, they missed her, too, they didn't want to take sides but they wanted Marilyn to stay in the church.

"It's a hard position for anyone to take sides," Marilyn said. "I mean, these are people you've known for years. To think that someone else in the church would do this to you, it blows them away, like it did us."

Dennis thought the church members were coming to Marilyn and trying to drive a wedge between them, telling her it was good of her to stay with Dennis, to stay married to a crazy husband.

How's Dennis doing? some of them asked. *We're really worried about him.*

Marilyn would think, *What about me? I'm in this, too.*

Once Julie Schaapman and Marilyn DeYoung saw each other. The two old friends hugged.

I wish this could all be straightened out, Julie said. *Someday it will be.*

Someday it will be, Marilyn agreed. She paused to think how that sounded, then she added, *But you know, I believe my husband.*

Julie said, *Well, I don't expect anything less than that.*

So Marilyn stuck with Dennis, and they both had many conversations with people to explain why they weren't going to church anymore. The Schaapmans were talking with a lot of people, too.

Dennis and Marilyn pulled their children out of Sunday school. It was a hard lesson. Eight-year-old Sara had friends there.

They tried another church for a while but didn't feel at home. And there were some people at the other church they thought were involved in waste-to-land schemes as well.

"It's like, everywhere you go—" Marilyn said, laughing. "I guess you know it's always going to be that way."

Marilyn had a beautician's license. She used to rent space in town and had planned to go back, but she couldn't face working in town because of all the gossip and negativity.

"We have been labeled 'the bankrupt farmers' by Cenex and all the people that support them, which are most of the farmers," Dennis said. "Cenex salespeople stop in at almost every farmer's land at least every other week. This is the perfect opportunity for Cenex to get their message out that I am a bankrupt farmer and deserved what I got. If Cenex wanted to, they could reach, and probably influence, every farmer in the county within a two-week span."

They were having dinner at a restaurant in Wenatchee when they saw their old Cenex field man, Dave Nerpel. In days past, they would have joined each other and talked sports, children, and religion. But Dennis believed Nerpel was part of the reason he'd lost his land. Nerpel had tipped off Dennis to the rinse pond fertilizer, but Dennis thought his friend acted too little, too late, after probably helping to arrange the deal in the first place.

When they spoke, it was all Dennis could do to contain his anger.

I know what happened. I know what you did.

Nerpel tried to calm him down. Dennis would not be pacified. The families separated and ate at their own tables.

———

The United States Department of Justice took a plea bargain from Cenex and John Williams in the spring of 1995, closing the criminal case that had started when EPA workers in space suits invaded the rinse pond site in Quincy.

The judgment was narrowly drawn. They pleaded guilty to using a pesticide for an unapproved purpose. The maximum penalty was $200,000. Cenex agreed to a $10,000 fine and agreed to give $3,000 worth of chemicals to government agencies. Williams's sentence was a $250 fine and one hundred hours of community service. Cenex and Williams were placed on a one-year probation.

Dennis was hoping for a better outcome to help his case on appeal, but he thought it was better than nothing. At least it was a conviction.

Patty was apoplectic. Before the judge approved the plea bargain, she sent him a fax on City of Quincy letterhead with a copy of the rinse pond test results on beryllium, cadmium, and chromium. Proof, she wrote, that the pond had far more toxic chemicals than the pesticides to which Cenex had admitted.

She wrote: "On behalf of the citizens of Quincy, both present and future, I am writing to urge you to impose the strictest possible penalty against Cenex-Full Circle for their illegal and highly irresponsible means of disposing of hazardous waste . . . PLEASE HOLD THEM ACCOUNTABLE. Thank you. Patricia Anne Martin, Mayor."

Ruthann Keith, the horse breeder, wrote the judge, too. Ruthann thought she'd been lied to by Williams and disrespected by the big shots at Cenex in Minneapolis. She got a few friends to sign the letter and sent it in.

The judge approved the plea bargain anyway.

Patty had hoped nobody in Quincy would see her fax. She was wrong. It was put in the court file for all to see.

Tony Gonzales was a council member, a coworker of Glenn Martin's, and a confidante of Patty's. He had known her since she was a teenager. But Tony would never forgive Patty for that fax. He was married to John Williams's daughter.

Gonzales, sturdy and full-faced, confronted Patty in the hallway at the town hall. *You're attacking* me *when you attack my father-in-law,* he said. She lost him as an ally forever.

The EPA investigators told Dennis and Patty they were embarrassed that so little came from a strong case. They said the United

States attorney in Yakima wanted the plea bargain. Later, the EPA refused to make public a copy of the investigators' case-closing memorandum.

Patty plunged into other work at the town hall. There was a budget to adopt, grants to write, a new swimming pool to build, and a downtown improvement district. Martin was stalling on a street-paving project. She didn't want to pave until she could be sure there wasn't any toxic waste in the asphalt.

Patty's mother was diagnosed with lung cancer. Patty put her on a diet regimen including shark cartilage. When Erika Naigle went through radiation and chemotherapy, she never lost her hair, never got sick, never had blood problems. Doctors didn't find any living cells in the tumor. Patty felt vindicated in her beliefs.

A short time later, Patty and Glenn rushed over to a hospital in Seattle because Glenn's mother was sick. She, too, was diagnosed with cancer. It was too late. Glenn's mother had tumors the size of half-dollars in her liver and died seven days later.

The fear of cancer motivated Patty. It was well known that farmers were at high risk of several types of cancer, possibly caused by pesticides, nitrates in water, or even solar radiation.[3] But nobody ever investigated the fertilizer.

CHAPTER 3

CONNECTIONS

DENNIS DROVE AROUND LOOKING FOR LOUSY CROPS in the Quincy Valley so that he could recruit other farmers in his fight.

He saw a poor field of wheat on Road 9 NW. Dry, stunted, spotty growth. He saw a heavy, dark-browed man out in the field changing sprinklers. He knew Tom Witte a little bit, but he'd never talked with him much. Dennis turned his truck around, parked, and walked out in the wheat field.

Tom, this wheat field looks like crap.

Yeah. Yeah. Witte paused and considered.

Looks just like mine, Dennis said with a cackle.

Yeah. Yeah.

You might want to check your fertilizer.

Dennis told Witte about his own problems, and Witte started to wonder about the fertilizer on his own poor fields. Although Witte hadn't used material from the rinse pond, he had used products from Cenex.

Tom Witte told Dennis he had just the thing. He pointed to a white steel fertilizer tank Cenex had delivered and set up on his property in 1991 and forgotten.

Dennis taped a tin can to a wood pole to reach in the tank and scoop out a sample. It was a dried-up, dusty residue of the liquid nitrogen fertilizer. He put it in a jar and sent it registered mail to an analytical laboratory in Idaho. Dennis couldn't afford the full screening, so he asked for metals he already suspected. The result came back.

Titanium: positive.

Chromium: positive.

So now Tom began to think it wasn't the weather or himself to blame for his poverty. Maybe it was Cenex. But he didn't know what to make of Dennis. Tom was used to doing things for himself.

Tom went over to his side yard, opened the hole on top of the tank, and scooped about two pounds of dust, rust, and residue from the bottom. He sent the material to a lab in Ohio for the full screen. A while later he got the results back.[1] As he looked down the list, he saw:

Arsenic, 36 parts per million.

Lead, 217 parts per million.

Holy Christ, Tom said.

He called Dennis. Dennis said. *They got you, too.*

They couldn't believe these toxic metals were in the fertilizer. They hadn't imagined looking for arsenic or lead. The two broke farmers thought they were holding the proof to solve the mystery—and the way out of their misfortune.

Now we know what to look for, Tom told Dennis. *We aren't just speculating and guessing.*

A short time later, Dennis gathered some crops that had been grown on fertilized ground and sent them to a testing laboratory. The lab found lead and arsenic in the peas, beans, Sudan grass hay, and potatoes. The potatoes tested highest for lead.

The circle closed: from waste to fertilizer to food.

A few miles away, a man who'd rarely passed up a fight or nearly anything else that looked like fun was having some crop problems of his own, and it didn't take much for Dennis to persuade him the fertilizer might be to blame.

His name was a mixed metaphor of cowboy and French legionnaire: Duke Giraud. Duke wore scuffed boots and a rumpled cowboy hat. He looked out from Coke-bottle glasses at the chemical companies near his produce company on the railroad track in Quincy, and seethed. Duke just knew they were trying to screw him over to take his family land. Everybody said Duke was the most likely one of the bunch to come out shooting. He was quick to anger and slow to forget.

The Girauds ran Quincy Farm Produce. They bought, packed, and sold potatoes and onions at wholesale. Duke's father had started the company in 1956. Now Duke ran the operations, and his wife, Jaycie, kept the books.

The Girauds also farmed. Potato and onion farms carried high risks and high rewards. In good times they were the most lucrative cash crops north of tobacco. They cost about two thousand dollars an acre to grow, more than most other crops, so the cost of fertilizer wasn't as important as it was to lower-valued crops where farmers had to watch every penny to make ends meet. Potato, onion, and tobacco farmers are some of the heaviest users of fertilizer.

The Girauds had survived in the business for forty years. Duke was farming four varieties of onions on sixty-five acres. In early 1992, he'd hired a sprayer to apply herbicide to his onion fields. It was supposed to kill weeds, but instead the onions died, starting the very next day. The weeds flourished. Duke knew you have to be careful with chemicals, but he thought there was something seriously wrong here.

Duke and his son were both developing asthma. Duke had his thyroid gland removed later in the year. He blamed the herbicides. He had no proof at all of a connection between the herbicides and his sickness, but he sued.

The chemical company field men were telling people Duke was a bad farmer. That made Jaycie and Duke madder than hell.

One day, Duke was having coffee with Dennis DeYoung at Paddy McGrew's restaurant, like they did once in a while, when Dennis told him about a new suspect he'd never thought about.

Check your fertilizer, Dennis said.

Duke thought Dennis was crazier than he was.

No, really, you don't know what's in the fertilizer, do you? Dennis said. *It could be anything. They dump anything in there and call it a product. That's how they get rid of their waste. That's how I lost my land.*

Duke started to think Dennis might have something there.

Quincy Farm Chemicals was suing Duke for nonpayment of his chemical bills. Now he started looking at the fertilizer, which was supposed to be pure. Duke got more and more involved with Dennis and Tom and Patty.

Duke and Jaycie had what some might call an outspoken relationship. They fought big-time over his emerging beliefs on hazardous waste in fertilizer. She didn't believe him. They'd be embarrassed to talk to some of their friends about the supposed toxic chemicals in fertilizer. They'd get a lot of blank stares.

It just can't happen, Jaycie said. *It's insane.*

Well, hell, I know it's insane, but it's happening all over, Duke said.

The Girauds sent some produce to a lab in Idaho to test for the arsenic and lead that Tom had found in his fertilizer tank.

"We couldn't afford to test for more," Duke said.

The results came back positive for lead.

POTATOES	4.3 PPM
BEANS	1.6 PPM
HAY	1.2 PPM

Lead causes brain damage and other ills, especially in children.

Duke and other growers in the Northwest make 80 percent of America's french-fried potatoes.

"I wonder if McDonald's knows about this," Duke said, grinning.

———

Tom Witte was reading everything he could get his hands on. *Soil Fertility: Renewal & Preservation; From the Soil Up; Mineral Nutrition of Higher Plants; Western Fertilizer Handbook.*

Tom tried to figure out where the toxic chemicals came from and where they went.

In his reading, he learned that landfill costs had gone up tenfold in the past ten years. The cost, insurance, liability, and environmental laws were all pushing heavy industry to keep its hazardous wastes out of landfills—to recycle and market them any way they could.

Tom learned that fertilizer was a vehicle of choice to dispose of wastes if they had a little fertilizer like zinc or iron, and a lot of toxic stuff. It seemed to be something of an open secret. It seemed to be— was it possible?—legal.

One day Tom called Megan White, director of the state office on hazardous waste and toxics reduction. White told him she worked with six hundred companies in Washington that needed to dispose of toxic wastes. If the wastes had agronomic benefits, putting them into fertilizer would indeed be a recommended method of disposal.

Six hundred companies. Waste to fertilizer. Recommended by the state.

Witte sat down and wrote a letter to the governor.

The reason for this letter is to ask you for help with a problem I and several other farmers have, and to explain to you some serious problems with the Depts. of Agriculture and Ecology.

The problem has to do with commercial fertilizer that has been adulterated with substances that contain various heavy metals. These substances are generally industrial wastes that are finding their way into commercial fertilizers that are being sold to unsuspecting farmers.

I am enclosing a copy of a test I had done on some residue found in the bottom of a liquid fertilizer tank I have in my possession. . . .

When I received the results of the test I gave a copy to Mr. DeYoung who then took samples of peas, beans, potatoes, and sudan grass hay to Silver Valley Laboratory in Idaho and had them analyzed for abnormal levels of arsenic and lead. These crops had all been grown on land that had been fertilized with liquid fertilizer from Cenex, Ltd., Quincy. All the samples came back with elevated levels of lead and arsenic. The potatoes were about ten times what is allowable. Who would ever think the French fried potatoes could be a source of lead poisoning?

We have to ask the question: How can these things happen in a state like Washington? We have laws. We have departments. We have regulators. We have investigators. We have standards. How can these things happen? How can farmers be ruined? How can consumers be poisoned? How can these things happen when we have all these safeguards in place? The simple answer is that nothing works like it is supposed to and that is really sad.

On Friday, Feb. 2nd, Dennis DeYoung, his father Jake DeYoung, and I spent the day at the Pesticide division USDA. We left Quincy at 5:00 AM, and spent the entire day looking at the file and talking to employees at the pesticide division. We were there for six hours and we learned a lot. We were favorably impressed by the people who work in the pesticide division. They were friendly, helpful, and concerned about our problems.

The most disturbing thing we found out, and the one thing you need to be made aware of is this. The state has no mechanism set up to prevent toxic heavy metals contamination of fertilizers. It is unlawful to adulterate fertilizer. The maximum penalty is a one thousand dollar fine. Fertilizer is only tested for fertility elements. Nobody checks on what is in the inert ingredients, so we have a situation tailor made for abuse. People in industry think that the best way to dispose of waste is to sell it for fertilizer and let unsuspecting farmers spread it on their land. People in industry have long believed that dilution is the solution to pollution, and what better way is there? Just spread the garbage on thousands and thousands of acres of farmland.

Sincerely,

Tom Witte

March 7, 1996

The governor never saw the letter. Instead, six weeks later, the agriculture director wrote. Tom read it closely to see if he might find an ally in a high place.

The letter said state agencies had already been working on "some of these interagency issues dealing with fertilizers and waste." Well, that wasn't reassuring, Tom thought, and it was an awfully bureaucratic way of talking about the wrong that he'd described. "The area of waste stream disposal, reuse and recycling is large and diverse."

No shit.

"I want to assure you that this and like waste by-products entering the fertilizer realm are being looked at seriously. We will be evaluating what is the best route for dealing with them over the next year."

Nothing. Nothing changes.

Tom drove over to John Williams's office at Cenex one day clutching the laboratory results on the residue in his fertilizer tank. Tom told him there were a lot of heavy metals there.

Williams didn't say much of anything. After Tom left, Williams told people he thought Tom was trying to get his hand in Cenex's pocket.

Williams had driven past the Witte farm many times. He could see the rusting equipment, the junk and weeds around the outbuildings. Those are never good signs. A neighbor who had taken over some land Tom abandoned after his wheat crop failed in the wreck of 1991 told Williams that Witte had been late getting water on, late planting seeds, and late spraying weeds. He said it was carelessness; Tom would say it was poverty.

"He is probably stretched way too thin to be a farmer," Williams said of Tom. "He doesn't have the equipment. He is farming on real marginal ground. He has got a real tough row to hoe even if he was doing a real good job of it."

Tom also sent a copy of the fertilizer test results, with arsenic and beryllium highlighted, by registered mail to the president of Cenex in Minneapolis and the U.S. attorney in Yakima and the head of the Farmers Home Administration in Ephrata. He did not hear back from those people.

———

"It's a public-health issue," Mayor Martin said, insisting on more testing at the Cenex site. Cenex wanted to dig up the topsoil and bury it in a hazardous-waste landfill to end the fear-mongering about microscopic poisons in the dust. Now the mayor stood in the way.

A council member pleaded with her, "Patty, just let them take the soil away."

"Destroying evidence," she replied. Patty told her husband, "The key to everything that's been applied in the Columbia Basin is in that soil at the rinsate pond."

As she emerged more into public light on the issue, Patty saw the effect of her activity, and it wasn't always good. Some farmers told her they were afraid they wouldn't be able to sell their crops. The business-people in town were more and more suspicious of the mayor's activity. The newspaper editor thought she was impossible to talk to. Every conversation turned to one subject.

One of the council members to whom Patty poured out her suspicions, Jess Slusher, a minister, said, *Geez, Patty, you need to get an investigative reporter on this.*

Patty talked with the Washington State Department of Ecology every week. She didn't know they were writing up her conversations and sharing information with the fertilizer industry lobbyists.

Her first clue to the mixed motives of state ecology officials came in a phone call. She was asking about evidence she'd found that a cement kiln incinerator in Seattle was getting rid of toxic ash by spreading it as agricultural lime.

Lime is commonly thought to be ground limestone, but it turned out to be so much more. Lime reduces the acidity of soil and helps loosen the soil, especially in areas with lots of rainfall. Patty chanced to see a public notice that talked about lime in a new and frightening context. It was a proposed state rule for disposing of a dangerous waste known as cement kiln dust as fertilizer.

She was quite sure the farmers wouldn't know what they were getting.

Cement kilns are the hottest of industrial processes, burning at three thousand degrees Fahrenheit, and for that reason are well suited as de facto waste incinerators. The kilns can burn hazardous waste, medical waste, anything, really, while making cement almost as an afterthought. The process leaves an ash known as cement kiln dust. Holnam Company ran a cement kiln near the Duwamish River in Seattle. The ash, collected from the smokestack, was 12.8 on the pH scale and classified as a dangerous waste.

The company said a material that alkaline would work as lime on soil. What Patty saw was something else: dioxins and furans—perhaps the most dangerous molecules known to man.[2] The cement kiln dust fertilizer was peppered with them.

A single exposure to dioxin molecules at a critical stage in fetal development can cause cancer fifty years later. Mimicking hormones, they cause birth defects and neurological damage. Dioxins are largely a twentieth-century industrial poison. By-products of high-temperature processes, they are captured by air pollution technology in the smokestacks of garbage incinerators, secondary metal smelters, and cement kilns.

Nobody imagined they'd be spread around in fertilizer.

Until now. Patty happened to see a notice showing Holnam shipping thirty thousand tons a year of cement kiln dust to Northern Lime Company of Burlington, a small town an hour's drive north of Seattle, to sell as a substitute for agricultural lime.

They needed a lot of lime on the rain-washed soils of coastal Washington State, but did they need this? Holnam argued it was an inexpensive and safe alternative to potassium fertilizers. The company also argued, à la the tobacco industry, that dioxins had not been proved to be unsafe.

But Patty learned that the EPA had, in fact, said farm families could get sick from dioxins if they ate the vegetables raised on fields limed with cement kiln dust and ate the beef and milk raised on feed from those fields.

But there was no proof any one individual had been hurt by cement kiln dust. The EPA used a mathematical model. Holnam said its material was cleaner than the dust from most of the 212 operating cement kilns in the United States, many of which had dioxin levels twice as high.

A state researcher had been working on the issue before Patty stumbled across it. The researcher, Bill Yake, thought there were legitimate health issues. He wanted to suspend the sales and check the sites to which the ash had been applied.

So when Patty talked with a state ecology department official about the report, she was surprised he did not seem interested in her views on using dioxin-laced ash on farm ground.

The official wanted to know one thing:

How did you find out about this?

It had been two years since Judge Cooper had thrown out Dennis DeYoung's case against Cenex. Now the state court of appeals tossed it back to him. A three-judge panel in Spokane unanimously reversed Cooper's rulings. They found credible evidence Dennis couldn't sell or lease the land because it was still damaged from the Cenex materials years later, when it was repossessed. That was the key issue in the case, suddenly alive again.

Dennis could not believe it. The court of appeals also overturned Cooper's ruling that Dennis owed Cenex $130,000 plus interest. The case was sent back to Grant County Superior Court. A jury of twelve citizens, not a lone judge, would decide whether Cenex's actions led to Dennis's loss.

In April, after Duke Giraud's onion crop was planted, Duke and Dennis drove to Spokane to the Department of Ecology office charged with managing toxic wastes in eastern Washington. Showing up unannounced, they told a receptionist they wanted to look at the files on a company called L-Bar. They were public records.

The receptionist called somebody in back. Duke and Dennis waited. Finally Jim Malm, regional manager of toxic waste reduction, came out and asked them what they wanted. The files. Malm went back to his office while Duke and Dennis sat in the waiting area. Malm returned and told them they should have called or written first. The files were open to the public of course, he said, but they didn't have an empty room where Duke and Dennis could sit to look at them.

Sorry. Please come back another time.

What was he thinking? Farmers are perhaps the most stubborn people on earth. Dennis and Duke refused to go.

We'll just look at them here, Duke said. *We'll just sit right here on the floor.*

Yeah, we don't need an office or a table. The floor is fine, Dennis said.

A few minutes later, Malm found them a table to sit at, left them a thick file, and Duke and Dennis, exchanging knowing looks, for they had won a small victory, began to page through the material.

They found the first proof beyond Cenex of contaminants in so-called fertilizer. It was a document talking about a lawsuit in Oregon complaining about the waste product sold as road deicer or—take your pick—fertilizer.

It was right there in the state file: solid evidence that farmers' fields were dying from the recycling practice while the largest aluminum company in the world was padding its profits.[3]

Northwest Alloys, a subsidiary of the Aluminum Company of America (Alcoa), ran a smelter in Addy, an hour's drive north of

Spokane. Between 1984 and 1992, Duke and Dennis read, the company recycled more than two hundred thousand tons of dangerous waste from the smelter through L-Bar, a much smaller company.

They sold the dangerous wastes as a fertilizer called Ag-Mag, CalMag, or AlMag, or a deicer called Road Clear. The same stuff. The chloride content that posed a risk as fertilizer made it effective as a deicer. Based on company-sponsored research that said the material was safe, state officials in Washington, Oregon, and Idaho had allowed it to be sold for both purposes.

All this was hidden in plain sight in the public files. Duke and Dennis shared a sense of outrage and vindication as they read on.

Alcoa had saved at least seventeen million dollars in disposal costs. Duke and Dennis learned that the raw material for CalMag was classified as a dangerous waste. It flunked the test for chlorides. By getting it classified as a product, the companies sidestepped the state rules on dangerous-waste storage.

Digging deep in the L-Bar file, Duke found a real interesting letter. It was from an extension agent for Oregon State University to a manager at L-Bar Products. He was writing in the spring of 1989, when L-Bar was trying furiously to sell or give away the stuff as fertilizer because it would cost so much to store it as a dangerous waste. The extension agent wrote that he would vouch for the material's usefulness as a source of potassium and magnesium and alkalinity for farm fields.

Then he added, plain as day in the letter to the company, "I want to thank you for your financial assistance."

Duke thought that sounded like L-Bar had bought itself a university scientist.

Later the scientist said he didn't remember how much L-Bar paid for his work, but he said it wasn't much and it went to the extension service, not him personally. The scientist, Gale Gingrich, said matter-of-factly, "A lot of researchers get funding from chemical and fertilizer companies." He'd heard about some of the crop problems from over-applying the L-Bar material, and he said he'd never concluded it would be safe in the environment, just that it would work as lime, if applied carefully, on bare ground.

Unfortunately it wasn't applied carefully. Soon the stuff was being spread all over the countryside. Many farmers seemed to get a good liming effect from the material, but others didn't. It turned out the L-Bar waste was not so safe, not even on bare ground.

Two Oregon farmers saw their clover, oat, and hay crops mysteriously wilt. They sued for seventy-one thousand dollars in losses and eight hundred thousand dollars in punitive damages. Their lawyer hired James Vomocil, an Oregon State University soils expert, to test the fields and fertilizers. In 1993, Vomocil said CalMag was unpredictable and unsafe and its sales brochure was "designed to deceive." With that, the farmers won out-of-court settlements. They declined to say how much.

So what did that have to do with Quincy?

Perhaps nothing, in the end. Cenex managers in Quincy and in its regional office claimed they never bought anything from L-Bar Products, never even heard of the company. Dennis and Duke found documents proving otherwise. They found a receipt in the file dated November 1, 1990, showing L-Bar Products sold at least thirty tons of "fertilizer" to the Cenex distributor in Quincy. It was proof they lied.

And who did Duke and Dennis find watching out for the public interest at the L-Bar waste site? Jim Malm, the same fellow in charge of the cleanup at International Titanium and Cenex. They found notes of a 1992 meeting at which Malm was told L-Bar needed to "find a market" for three thousand tons of waste before Alcoa headquarters would finance a plant expansion project.

Another document showed Cenex was considering buying thirty thousand tons. L-Bar later said it sold small amounts to Cenex in Quincy but never made the big sale. The timing matched the farm troubles. Duke and Dennis were sure some of the CalMag wound up on fields in Quincy.

They just couldn't prove it.

Driving home, Duke and Dennis talked it all over, no longer thinking just of themselves and their failures in court, but of the bigger picture. The sun was low to the west when they passed the former site of International Titanium outside Moses Lake.

A few days later, Duke phoned the state capital and asked to talk to the attorney general herself. Christine Gregoire was developing a national reputation fighting tobacco companies; she was also previously the director of the state department of ecology, helping companies recycle dangerous wastes into fertilizers. Duke was transferred to an assistant of Gregoire. As he recalled the conversation:

Is it legal to put hazardous waste into fertilizer? Duke asked.

How'd you get my phone number? the assistant answered.

I just want to know if it's legal to put hazardous waste into fertilizer, or not.

How'd you get my phone number?

The attorney general's office was no help.

Duke told Jaycie later, *The more I dig into this, it just becomes like an obsession because I can't believe it's going on.*

Tom Witte had a sister named Nancy living in Fairbanks, Alaska. Nancy Witte was a special-education teacher and occupational therapist with twenty-five years' experience with disabled children. She wondered why so many kids were sick. The rate of autism had tripled in fifteen years to fifteen per one thousand births; Down's syndrome children were born to younger and younger mothers; and attention deficit disorder was being diagnosed, and treated with Ritalin, everywhere.

Nancy Witte was also a basketball player, a bit older and a lot smaller and just as fierce as Patty Martin.

So it was only natural that when Nancy came back to Quincy to help on the farm, taking a year's leave from the school district, she would team up with Patty.

She's the best friend we've got, Dennis told Nancy. *She needs help. You're strong. You're someone who can help the mayor.*

They didn't know each other. The Wittes had gone to school in Ephrata, twenty miles south of the Martin clan in Quincy. Nancy talked with Patty on the phone a few times. They hit it off. They wanted to meet. But Patty didn't want the Wittes to come to her house because she thought it was being watched.

Patty suggested they meet at a public event called Family Night at the Park. There were going to be three hundred people at the town barbecue. Nancy and Tom didn't know what that was all about. They didn't understand Patty's cautiousness, bordering on paranoia, until later. They just drove to the barbecue, dressed a little better than they usually were after a day's work, on a hot August evening.

This is an event we would never *go to,* Nancy said as they walked up to a small crowd gathered in the swath of green grass and shade trees bordered by the West Canal.

The Wittes recognized nobody, and nobody recognized them. There were not three hundred people at the park, but seventy-five. Everybody could see everybody and hear half of what they were saying.

Tom and Nancy checked the food. Hot dogs and potato chips and soda pop. Grinning, Tom took a hot dog from the grill. There was nothing there that Nancy would eat, so she walked up to a woman she didn't know and asked: *Who's Patty Martin?*

The woman pointed out the statuesque mayor. Tom and Nancy ambled up to Patty and said hello. It was not a good first impression.

I can't talk now. Patty looked away.

She saw Tony Gonzales nearby. He was serving popcorn from a Cenex popcorn popper. Patty thought Tony might recognize Tom. A lot of the other people Patty thought were involved in toxic-waste schemes were in the park, too. She thought they might be watching her and the Wittes.

Patty turned away.

What the hell's going on? Nancy whispered to Tom. *Why's she so afraid of being seen with us?*

Tom and Nancy walked off and sat in the shade on the grass part-way up a bank, looking over the picnickers in the sun and a band on a stage. A few minutes later, Patty came over and sat down on the lawn about eight feet away. Patty kept looking over her shoulder to see if she was being watched. She kept her eyes on the band. She talked out of the side of her mouth like a character in a bad spy movie. It looked to Nancy, who wasn't yet aware of the mayor's paranoia, like Patty was scowling at them. Nancy knew that even in rural America there are classes of people, and she wondered if Patty, a middle-class townie, wanted anything to do with poor farmers like her brother.

—

Dr. Max Hammond phoned Patty one day and arranged to meet her at the town hall. He'd heard the mayor was being overspeculative.

Hammond held a bachelor's degree in agronomy, a master's in soil microbiology, and a Ph.D. in plant science from Utah State University. Patty thought Hammond was overqualified for the job of Cenex scientist in the Columbia Basin. She suspected, without a shred of evidence, that he might be the mastermind of waste-to-fertilizer in all of Cenex/Land O'Lakes.

Patty asked the city attorney whether she should ask Dr. Max pointed questions. He said she would learn more if she just listened. Patty took four pages of notes on the conversation.

The pond didn't work, Hammond told her. *It was built to solve a problem, but it didn't work.* Cenex emptied and buried the pond after it failed to evaporate like it was supposed to. Hammond said he checked Dennis's land before and after the rinse pond material was put down there. Patty asked what was in the pond. *Pesticide residues,* Hammond said. But the pond was never registered to hold pesticides.

That was a mistake, he said.

—

Max Hammond emerged as the contact person with regulators and the concerned public on the Cenex site. The company dug five wells around the old rinse pond after the invasion of the EPA. They promised to check the water for signs of contamination.

One day Hammond stood at the front of a public meeting room to release the groundwater tests. The room was almost empty. Mayor Martin was there, and on the other side, Tom and Nancy Witte and Duke Giraud sitting together. Two state officials were there as well, and a university scientist invited by Mayor Martin.

They were tracking a plume. Hammond pointed to charts and maps and tables on an overhead projector screen. He said Cenex would clean up the pollution, especially the 1, 2-dichloropropane. The state officials nodded.

Then Patty stood up. She had an overhead transparency of her own. It was a copy of a page Duke had found in files at the state capital, test results from the Cenex soil.

BERYLLIUM	227 PPM
CADMIUM	8 PPM
CHROMIUM	181 PPM

I'd like to know about these metals, Patty said.

These were very high levels of toxic metals. Hammond hadn't mentioned any numbers like these. The state officials looked surprised. From then, Patty thought, the meeting turned from Max's agenda to hers.

It turned from a cool discussion of 1, 2-dichloropropane to a flaming debate about God-knows-what, because nobody on earth knew what was really in the material Cenex had called a fertilizer.

The chemical poisons that had spilled or leaked from the Cenex pond had trickled through ten feet of soil and pooled on a ten-foot layer of impermeable rock. It wasn't going anyplace. Now Patty said the Cenex charts showed one of the monitoring wells had been tapped through the lowest point of the rock into the lowest aquifer tested. In short, she thought the monitoring well was a sink allowing the poisons to drain into the aquifer.

They've inoculated the aquifer, Patty said.

And she thought but did not say, What were they *really* trying to get rid of?

Soon the meeting broke up. The university scientist who had been invited by Patty agreed it appeared the monitoring well had tapped the aquifer.

So it was out in the open. Patty and the broke farmers were coming out. They were going to push—and they expected to be pushed back.

The end of this meeting marked the beginning of the water group, so named because they'd believed toxics from fertilizer were seeping into well water. These were the few citizens—outcasts, mostly; the Girauds, Wittes, Dennis DeYoung, and Patty Martin—trying to figure it out. They met in the lobby area after the meeting for a few minutes. The group decided to get together at the Girauds' house the next weekend.

Patty and Dennis thought they were in danger. Only a handful of people knew about the toxic-waste-to-fertilizer connection. They needed to share information. They needed to protect each other.

——

Patty told her husband, no one else, where she was going.

She drove a mile north of town, turned right at Road 6, crossed an irrigation canal, and turned right at a dirt driveway. The yard was screened from view of the road by two rows of trees rustling in a summer breeze. She passed a low-lying house of sixties suburban design; cheap brick, tar roof, and unframed windows. Patty drove around back. There was a small yard of dry, yellowed grass, enclosed by a chain-link fence for dogs.

They were waiting for her on the cement patio in the shade. Duke, Jaycie, Dennis, Tom, Nancy, Russ: the water group was meeting.

Duke had been two years behind Patty Naigle in high school and was now thirty pounds heavier. Patty had hardly aged; at her twentieth high-school reunion, she'd been voted the person who had changed the least. Duke thought she was smart, determined, and genuinely concerned about the health of people who were living among all those chemicals.

The group met over a half dozen times, always on a weekend,

always at Duke's secluded house. They came to believe, as Witte put it, "All the agencies involved know what is going on and their main job is to keep it covered up. . . . What we have here is an industry that is for all practical purposes unregulated. We have a system that invites and rewards fraudulent behavior. We get wastes from everywhere. We get wastes from pulp mills, mining wastes, incinerated medical wastes, and aluminum industry wastes, just to name a few sources. From Canada we get fertilizer made from scrap batteries."

Dennis wondered why any beryllium, cadmium, and chromium were present at all in the Cenex fertilizer rinse pond. He drove to the Quincy library and thumbed through the Merck index of chemicals and drugs.

Under beryllium: "Death may result from extremely low concentrations of the element and its salts." Cadmium: "Ingestion causes choking, abdominal pain, vomiting, diarrhea. Inhalation causes cough, headache, vomiting, chest pains, pneumonitas, bronchopneumonia." Chromium: "Chromium and its compounds are extremely toxic."

What are we eating? Nancy asked. *What's happening to us?*

Duke and Jaycie talked about their kids' respiratory problems and asthma and allergies. Duke wondered if he'd eaten something that was affecting his heart and lungs. Duke had quit smoking. His lungs still hurt. Jaycie wasn't feeling well, either, but she still smoked.

Some people in town blamed the Girauds' smoking for their children's lung problems. But the water group blamed the toxic chemicals.

A lot of people have idiopathic pulmonary scarring, Patty said. *There's a tie to this area.*

Oh, without a doubt, Dennis said.

Yeah, Nancy said. *It's something here.*

⟡

Patty had inherited an administrative assistant from Mayor Adams. One day the assistant quit, saying the work was unbearable because Mayor Martin didn't trust her.

Patty thought she was being set up. She thought Cenex was

inducing the employee's complaints, and of course Cenex wanted to discredit the work Patty did on toxic waste and fertilizer.

So she phoned the state capital and talked to a man whose job it was to help mayors. She'd met him in a class for the newly elected. He told Patty about a mayor who'd challenged the racist white supremacists in his small town. That mayor got help and protection from a newspaper. Would Patty talk to a journalist?

Yes, this needs to get out. We need help.

So the man spoke with his aide, who phoned a reporter she knew, who phoned me. Something about toxic wastes in fertilizer. Sounded like something I'd like to work on. Sounded far-fetched, but who knows?

CHAPTER 4

DIGGING

IT WOULD NOT BE INACCURATE to call me a muckraker. Nor would I be offended. The great journalist Lincoln Steffens applied the phrase to his probing accounts of urban poverty, corruption, and decay in the early 1900s. I'd taken college classes from Steffens's son, and I'd come to adopt another venerable journalistic motto as my professional purpose: to comfort the afflicted, and afflict the comfortable.

They weren't goals I could fulfill in a weekly newspaper like the one my parents ran in the small town I'd left after high school. I escaped to Western Washington University and then the Columbia University Graduate School of Journalism in New York City. I settled in Seattle, close enough to home. By now I'd been an investigative reporter at Seattle's daily newspapers for fifteen years. I liked the city life. A city has size and scope, and the protection of anonymity as well.

One of my investigations caused a juvenile-court judge whom I had exposed as a pedophile to commit suicide. (I didn't know he was suicidal. But I wouldn't drop the story when he offered to leave the bench and move to California.) My recent investigations had covered the deaths of 104 children under the care of the state; the deaths of four firefighters in an arson set by a spoiled frozen-food heir who escaped to Brazil; and the suicide of rock star Kurt Cobain. They were big stories in the northwestern United States, and they took me to the boulevards of Los Angeles, the barrios of Rio de Janeiro, and other places far more exciting than the fields and orchards I had left behind.

My subjects, at their most basic, were sex, death, and disaster. I was the kind of reporter that governors, senators, and especially, in my case, judges, didn't like to hear on the other end of the telephone line.

And to tell the truth, a lot of other people apparently didn't think much of my work, either. Public opinion places reporters right down there with lawyers, panhandlers, and politicians.

So maybe Martin and I had more in common than I thought; after all, she was a politician. I was suspicious of her from the start. Almost every politician who had ever wanted to talk off the record was hiding self-interest and personal motives. No politician who had leaked me information was unaware of the risks, while this woman sitting across my table seemed to be. She may have been suspicious and paranoid, but Patty Martin also seemed quick to trust, and a bit naive.

We met in the lounge of a hotel near Seattle. Martin drank ice water and shredded her napkin as she spoke. Duke Giraud, a rumpled man in a cowboy hat, sat with her. The story came out in a rush, an allegation overload. I like to let people ramble, and listen. My questions come at the end.

Martin said she thought she could be killed for talking about this. Somebody was listening in on her phone. Somebody was spying on her house. Her aide was quitting in an effort to discredit her. They all had connections to the chemical industry that dominated Quincy.

Mayor Martin talked about some farmers, Duke, Dennis, and Tom, who had spent four years snooping around to try to figure out why they were sick and their crops were failing. They thought they'd been singled out for dumping because they were already poor.

I could tell Martin would eventually go on the record. She was helping broke farmers fight the chemical companies—it had potential for a news story. And she was, after all, the mayor. In the news business, that planted the hook.

It took Martin an hour to sketch it out for me. I'd written about Hanford before, but never about food or farming. I drove back to my office, sat down at my computer, and wrote up the highlights, fresh in my mind, on a single piece of paper.

David Boardman was the paper's prize editor. Dave was smart, ambitious, aggressive, and, above all, busy, much sought for his news judgment and phrase making. He was a Northwestern University journalism graduate who wrote funny songs for retirement parties and sang them a cappella. People always marveled at his excellent tenor voice.

I told Dave what I'd heard in two sentences: Toxic wastes are being used as fertilizers. The government performs no tests and sets no limits and requires no disclosure to buyers for the heavy metals hidden in plant food.

That seemed like news to me. It seemed important. But it also seemed unbelievable. We're professional skeptics. My own first reaction had been to think Patty Martin had her facts wrong, or that she misunderstood. Dave said he'd give me the time to dig into it. If just half of what she said was true, it would still be a hell of a story.

My job was to separate fact from suspicion and put Quincy in a broader context.

I drove over the Cascade mountains, across the Columbia River, and up to the Quincy Valley, where farmers posted the names of their crops on signs along the highway.

QUINCY, WASHINGTON

Duke said he'd show me the proof. He'd give me the airport tour.

We climbed into the four-wheel-drive with the license plate DUKE1 and buckled up. Duke sped to a scrubland twenty miles east of Quincy.

It was the industrial area by the Ephrata Airport, a stretch of asphalt and scattered aluminum warehouses by a rail line. The airport terminal was a small, lost building. Not a single train, plane, or truck was moving on the bleached terrain, but in my mind's eye they could all come together here. Duke said he could imagine a lot of dirty dealing. He said this was the nexus of the hazardous-waste-to-fertilizer trade in the Columbia River Basin. I wondered if he was crazy.

He pointed to a heavy barrel-shaped object rusting in an empty field across the street from the little airport terminal. It looked like an oversize cement mixer, discarded, owned by no one. The machine had only one purpose: to roll slurry into granules like fertilizer.

He drove me to the warehouses where L-Bar, the Alcoa waste handler, stored its combination fertilizer/road deicer. We walked around the three big windowless sheds. We peered inside a sliding door that happened to be unlocked facing the rail line. The warehouse was full to overflowing with gray pebbly material. The air smelled of ammonia, and I wondered what I was breathing.

If they can't market it as fertilizer and if they can't give it away as road deicer, Duke said, they'll have to dispose of it as a dangerous waste.

Later I checked that. It was true. The man who devised the fertilizer plan for Alcoa waste, Frank Melfi, told me it saved the company two million dollars a year in landfill costs.

Somebody in a white pickup truck pulled out of a parking lot and followed Duke's truck as he drove along the asphalt roads crisscrossing the scrubland.

Duke talked as he drove. "What they're doing is using agriculture as part of this waste stream. The sheer volume of agricultural fertilizer would make it appealing, because if you did it right, you could spread it everywhere. Instead of paying two hundred dollars a ton to get rid of it, you could sell it."

Later I checked. True again. Many companies, especially in secondary metals, saved millions of dollars on toxic-waste disposal by paying fifty or one hundred dollars a ton to use it as fertilizer instead of two hundred or three hundred dollars a ton to store it in a landfill or recycle it safely.

"I'd like to find out who is doing this, if it's everybody, or they're just renegades." Duke watched the white truck in the rearview mirror.

Later I checked. They were not renegades. They were government-sanctioned businesspeople. It was legal to spread hazardous wastes as fertilizer, liming material, or soil supplements. Almost everybody in the industry was doing it in one way or another. They called it beneficial recycling.

"If this is legal and so good," Duke wondered, "why are they keeping it secret from everybody? Why doesn't anybody know? Nobody knows. Nobody'd buy it if they knew."

I asked him what the authorities ought to do. "They have to start testing fertilizer for what they *don't* say is in there, because they have no problem letting them add who-knows-what."

Duke said many times, "When you tell people this, they think you're nuts."

But he didn't sound nuts to me anymore. Mad, yes. Nuts, no.

After Duke took me back to town, I drove my own car over to Max Hammond's house, a rambler on a hill in green fields outside Quincy. "Dr. Max" was the top scientist for Cenex and the oldest male in a longtime Quincy clan.

Hammond coughed and choked as he tried to talk—*Interstitial fibrosis, scarring of the lungs*, he said, grimacing. His doctors could not identify the cause. His face was puffy from steroids for lung inflammation.

He settled in a chair by the window in the living room. His wife worried in the kitchen. Hammond had a bottle of oxygen by the chair but never brought the plastic hose, his lifeline, to his nose as he talked about the Cenex rinse pond.

Hammond said he knew the DeYoung case made some people think

Cenex was sneaking toxic waste into fertilizer, but he didn't think so. Cenex had spent millions cleaning up messes from companies it acquired. This was a misunderstanding. Cenex had a good reputation. He would talk to the Fertilizer Institute in Washington, D.C., and they'd support him.

Hammond labored for breath as he continued. "Occasionally you'll find industrial by-products that will be utilized as fertilizers. A lot of it comes from the mining or metals industries. But they have to go through some tests, probably very rigorous tests, to make sure you don't have a sleeper in there."

He praised Martin for watchdogging but questioned her for never letting go. Hammond said he was perplexed Martin didn't accept a second set of test results from the Cenex pond showing it was fine, after the first set showed high levels of toxic poisons. "Of all the things to go wrong—of all the dumb luck—the metals analysis is off," he said. "But even though we have rerun tests, they don't accept it. The health department understands. Ecology understands. Unfortunately there are some folks in Quincy who don't. I think we're reaching a point where perception outweighs the facts."

But what about the lab analysis of Tom Witte's fertilizer? I asked. The report showed toxic heavy metals no one would feed to a plant. Many of the same toxics found in the Cenex rinse pond were also found in the Cenex fertilizer tank on Witte's farm.

Hammond studied the lab report. Some heavy metals you'd expect to find from an old storage tank—aluminum in the metal, chromium in the weld, lead in the paint, he said.

But he was intrigued by the beryllium. He didn't know why beryllium would be in a Cenex fertilizer tank. He wondered about the cadmium, too, "and the arsenic, I just don't have an answer for that."

He would look into it. He would let me know.

Max Hammond struggled out of his chair and walked carefully to the front door to see me out.

"I feel like roadkill," he said.

I drove back to Seattle in the dark. I needed to finish another story before starting this one.

Hammond died five weeks later.

OCTOBER 1996 — QUINCY, WASHINGTON

The water group's biggest meeting was the day they gave their hair. Nancy Witte had contacted a homeopath in Seattle who suggested hair analysis to trace the heavy metals that had built up in deep tissues. Growing at about one inch per month—slower on children—hair samples, like tree rings, provide a long-term record of environmental exposure. The Centers for Disease Control questions the validity, but the EPA says there is a consensus of specialists that human hair samples, carefully collected and analyzed, provide good data.

The Quincy group jumped on the idea. The homeopath, David Vaughn, drove his red sports car from Seattle to Quincy to take the samples. Vaughn felt right at home. He'd grown up around horses in Bozeman, Montana.

Nancy Witte met Vaughn at a gas station and led him to the Witte dairy. Vaughn clipped some hairs from a cow and two calves. Each time he handled the hair with plastic gloves, measured it on a little scale, put it in a plastic envelope, and filled out a form.

They trooped over to Duke's house. The whole Witte family joined them: Tom and Nancy's mother, Agnes, Tom's wife Terri, their boys Sean and Bryan, Tom and Nancy's brothers Warren and Rock, and Rock's wife Cindy. Rock Witte was the center of attention because he was the sickest person in the room. Rock had been almost nowhere besides Grant County, and he had really worked with the soil. He did the chemical spreading and disking.

"He was having massive chest spasms and neurological symptoms and he was going to heart clinics all over the place, including the one over in Spokane, and not getting *any* help whatsoever," Nancy explained. "We thought this was our last chance to get him some help."

Dennis and Marilyn showed up, and Russ Sligar and his daughter and granddaughter, and a junior-high-school teacher with lung trouble. Duke and Jaycie served water and soda to the hopeful group of two dozen. Some people think homeopaths are quacks; these people wanted to trust. They'd dressed in their Sunday best.

We've never looked this *good before,* Dennis said, laughing. *I don't think I've ever seen us this clean.*

Vaughn identified himself as a doctor of homeopathy, a geologist, an orthomolecular specialist, and a nutritionist. He was not a nationally renowned research doctor of the type Patty and Nancy would have liked to help, but that type had not returned their phone calls. Vaughn was enthused. He was concerned. A one-time uranium miner, he had studied at both naturopathic school and medical school.

Vaughn cut each person's hair by measuring along a line on the center of the back of the head from ear to ear, traversing the occipital, or rear, lobe of the brain. He cut the hair close to the scalp. The kids grimaced. The adults grinned. They all felt more than a little like lab rats.

Duke held his dog, Bingo, while Vaughn cut his hair, too. Everybody who was concerned had showed up, but one: Patty. Nancy took Vaughn over to the Martins' house, where they took some hair from Patty and her youngest child, Eric.

Eric's hair was supposed to be a control sample. The child's body was supposed to be clean.

CHICAGO, ILLINOIS

Doctor's Data had never seen data like that from the hair samples of people and animals in Quincy, Washington.

Aluminum, arsenic, cadmium, lead, mercury, uranium—all of them unnaturally elevated. Something was wrong. The lab had to stake its good name on the findings. Doctor's Data had twenty-five years' experience in environmental toxicity and forty-five corporate clients. The technician checked the lab equipment and reran the samples. Everything checked out.

"High lead and cadmium like this would be found in a welder," the head biochemist said. "When you start seeing these things in children you know aren't welding, you know they're coming from the environment."

This couldn't wait for the mail.

The biochemist at Doctor's Data phoned the homeopath in Seattle. Some of the Quincy people had such high levels of heavy metals, he warned, they might soon die.

QUINCY, WASHINGTON

The homeopath called the Wittes first because they'd had the most creatures tested: three adults, two children, a calf, and a dog.

Bad news, he said. *I've seen hundreds of tests like this, but none this bad.*

Nancy Witte called Patty Martin.

It's our worst fears come true, the mayor said. She asked, warily, *How was my son?*

Nancy didn't know.

A few days later, the rural mail carrier pushed the stack of laboratory reports into the Witte mailbox. Tom and Nancy had finished milking, feeding, and cleaning the cows for the day. Tom pulled the envelope out of the mailbox and went inside to open it. They sat at a dining table, cluttered with detritus of two children, and started to page through the reports.

They were startled by the hair analysis of the children. The thirteen-month-old granddaughter of a friend of Dennis's had the highest level of mercury in the group. The expression "mad as a hatter" came from hat makers poisoned by mercury salts. Mercury traces, often found in fish, can cause neurological damage.

Tom's son Bryan showed beryllium, a residue of atomic and aerospace industries, in his hair. Beryllium poisoning is an incurable disease marked by scarring of the lungs. Beryllium was used at the Hanford Nuclear Reservation from 1952 to 1987 to close the end of uranium fuel rods. Hanford workers formed a Beryllium Awareness Group. But Tom's five-year-old had never been to Hanford. Quincy was the only place he'd ever lived.

This is amazing, Nancy said, *but somehow it's not surprising.*

Tom and Nancy compared test reports side by side on the table. Even a homeopathy non-believer (like me) could be worried by this. The Wittes were believers. They were scared by the lab results. Tom's and the boys' hair tests showed a pattern they came to know as the Quincy pattern, while relatives from outside the area showed different chemicals in their hair. Their mother's hair looked the best; she hadn't lived in Quincy in a while and didn't have much mercury or beryllium. Nancy had uranium in her hair, but she'd moved around a lot and the uranium could have been absorbed in New Mexico, Alaska, or Washington.

Tom and Nancy drove over to Dennis's mill to show him the results. Dennis's uranium exposure was almost off the chart.

You should be a lightbulb, Nancy told him.

Yeah, I'm the brightest one of the bunch, Dennis cackled.

He slowly read the list of toxins. It was easy to connect the results in hair to what seemed similar results in the Cenex pond. So that's what I was disking in, Dennis said.

Dennis took the lab reports over to Patty. She had expected to find the metals in farm families but not in her five-year-old son.

Oh no, Patty said. *He's got it, too.*

Eric's hair was high in aluminum, cadmium, lead, mercury, selenium, and uranium.

But the boy seemed perfectly healthy. Maybe less easygoing than

Patty's other children. Maybe a little aggressive. Sometimes, if she told herself the truth, very aggressive. As Patty thought about it, she wondered how heavy metals might be affecting her smallest son.

They all wondered. Nobody in the water group was dying of cancer. They all had vague, diffuse symptoms that could be associated with a variety of causes.

The homeopath recommended chelation for everybody. Chelation is a controversial course of treatment that claims to cleanse toxic metals, particularly lead, from the body by injecting a complex molecule that binds with metals and removes them through urine.

Patty didn't want her five-year-old chelated. It was scary to think of the boy hooked to a needle.

The farmers might have tried it but couldn't afford it so the homeopath told them to eat beans and garlic and diet supplements to pull the toxic metals out of their bodies.

Rock Witte was a harder case. Tom's younger brother was only forty-one but had suffered heart, lung, and nerve problems for the past decade. He'd been treated by doctors at a heart institute who couldn't pinpoint the cause. His test results were the worst of the bunch. He had lived in the Quincy area his whole life. Eventually he would try chelation, and after fifteen treatments, he said he felt a whole lot better.

Nancy and Patty slipped into the town hall to make copies of the Doctor's Data reports. They were glad nobody was around on the Saturday morning to ask them what they were doing. Patty was so nervous. They went in quietly to run copies of paperwork they believed would explode the practice of mixing hazardous waste into fertilizer, recycling dangerous chemicals through food and through people.

When they were done, Patty dropped two dollar bills on the clerk's desk to pay for the copies, a dime a page.

The next day, they all gathered for another meeting at Duke's house. Somebody new wanted to join the ragged band of broke farmers and crusading mayor who called themselves the water group.

Patty thought he was a spy. He looked like an old hippie, like Jerry Garcia with his long gray hair, unkempt beard, and floppy cotton hat.

Bill Weiss was in his mid-fifties and ran an organic farm on Road 7 NW a few miles out of town. He was the only person Duke knew who grew organic onions. You have to be certified pesticide-free for three years to qualify as an organic farmer. Duke thought Bill was like the other characters who gathered in his backyard—intelligent, independent, and sometimes isolated.

Weiss was friendly but quiet, a careful listener, and he took notes. Weiss told the group he had an idea of what they were up to. Dennis had talked with him, and he was curious to learn more. He had already asked his own tenants at the organic farm if they were using recycled waste in their fertilizer—only to be greeted a few hours later by the two big guns from the fertilizer industry in Quincy. Max Hammond and John Williams had shown up at Weiss's farm to assure him there was no such thing going into fertilizer.

Naturally, that made him more suspicious.

Bill Weiss knew Dr. Max well. They'd gone to Quincy High School together. "There's an underlying attitude in chemical agriculture, that people become true believers in chemicals," Bill Weiss said later. "It becomes a religious issue."

Bill said he was troubled by the closeness of the Quincy chemical plants to the schools. And he was outraged if in fact fertilizer companies were adding heavy metals to the soil. His son was a lawyer, and he'd be digging further.

Patty's suspicions lingered. Bill Weiss took such careful notes. She wondered why.

It was a warm fall day as the water group sat on the concrete patio out back of Duke's house, whiffing chemicals from a nearby field and talking about toxic waste in fertilizer. They decided that by working with the reporter and writing letters, they would make a record of the problems in Quincy and protect each other.

Duke wrote to a toxic waste reduction manager for the state

department of ecology. Duke did not like this man. He was the one who'd been less than helpful when Dennis and Duke had wanted to look through the files. Duke thought the man was more interested in helping industries reduce toxic wastes by plowing them into the ground. He took it personally.

Duke had no use for bureaucrats with fancy suits and hidden agendas. They should serve him. They could sure spend his tax money. Today he asked for a list of companies that used toxic waste in their products.

"I feel that the public (me included) has been very misled by your department not letting them know, in fact hiding the fact that toxic waste is going into their food," Duke wrote. "What's next, L-Bar on your corn flakes in the mornings? Your department has not only been no help to the public but I feel that you are trying to hide something from us. Every time we ask questions we get evasive answers and are talked to like we are brain dead."

Duke thought he'd get more evasive answers, but the bureaucrat would surprise him.

Jaycie Giraud sent an E-mail to a medical service she'd found on the Internet. "My husband, myself and three children all have a long list of heavy metals in our bodies," Jaycie wrote, blaming toxics in the water and dust. "My husband and a 13-year-old son have lung problems and some unexplained scarring in their lungs."

A nurse from Metal Detox Medical Services wrote back, "Your story is familiar—this type of thing is happening all over. In some instances, it seems to be somewhat of an epidemic. . . . Particularly from pesticides (arsenic) and other environmental pollution associated with farming communities (mercury, cadmium, lead, aluminum). With these particular metals, respiratory problems are common. Some of the damage is permanent. Some of the damage can be corrected. Do consider researching this area for your family's (if not community's) protection."

But the Washington State Department of Health was far more skeptical. The hair test results were inconclusive, a state toxicologist told

Martin. They failed to show a pattern of exposure to a specific toxin. They were all over the place. Further, the toxicologist added, the Centers for Disease Control did not find hair analysis useful in detecting environmental exposures or measuring most metals in the body.

Patty already knew the CDC was down on hair tests. She knew other responsible agencies, including the EPA, recommended hair as a good indicator of some exposures.

The state toxicologist also dismissed Patty's complaint about cancers in the Quincy area. She told Patty there was no unusual pattern of people dying from any particular cause in Grant County, which included Quincy. Later Patty learned the state only tracked deaths, not illnesses, and didn't count deaths of Grant County citizens who spent their final days in hospitals outside the county, as many did.

The toxicologist said local health departments might do more testing in Quincy. It never happened. The toxicologist said Martin was impossible to satisfy.

CHAPTER 5

LEAD IN YOUR FRENCH FRIES?

NOVEMBER 1996 — QUINCY, WASHINGTON

THE ELM AND HONEY LOCUST shed their leaves on the Witte farm. Winter blew in early and hard.

Tom and Nancy Witte and two hired hands worked dawn to dark. It was pure cold. For a while, they had a storm blow through every twelve to thirty-six hours with high winds and a foot of new snow. They used a space heater to thaw out fuel lines in the farm equipment. They nailed up plywood to keep the snow out of calf hutches. They had thirty-four heifers kicking the hell out of the milking parlor.

As they worked, Tom and his sister Nancy almost always talked about the damage that fertilizer with toxic waste might be doing. They came to some basic understandings. They thought they were figuring out something important the American farmer ought to know.

Tom wrote it down. Late at night and early in the morning, Tom sat at his dining table and wrote his thoughts in longhand on a yellow pad.

The next night, Nancy would type and edit what Tom had written. Their mother, Agnes, and Tom's wife, Terri, read it over.

While other farmers sharpened the edges of their hoes and axes, Tom honed his rhetoric. One of his favorite sayings of the time was, "The modus operandi is to lie and deny." And how he loved to chuckle at the government people who said they could spread industry wastes at "agronomic rates"—that was a euphemism worthy of *Animal Farm*. Agronomic rates indeed. It meant a dose that wouldn't kill the plant.

Tom and Nancy thought a lot about the title of the manifesto they were writing. It had to sound the alarm.

They thought about "Mercury in Your Milk" because they'd found mercury in Tom's fertilizer tank at five times the background level and they found it in calf hair, but they didn't have any tests showing mercury in milk itself. They were told milk would gum up the testing equipment. Later, it occurred to some people in Quincy that the Wittes might not want to talk about mercury in milk anyway; after all, they owned a dairy.

They thought about "Thallium in Your Toast" because they'd found 6.7 parts per million of thallium in the Cenex rinse pond, but there were no tests showing thallium in wheat, bread, or people.

So they settled on "Lead in Your French Fries."

After all, they had proof of high levels of lead in fertilizer, in Duke's potatoes, and in people. While lead was a natural element, it was also a well-known toxin. The government had banned lead in gasoline, paint, and food-can solder, but there was no limit on lead in fertilizer.

Perfect, Tom and Nancy thought. At the last hour they decided to add a question mark to the title "Lead in Your French Fries?" because while they'd found lead in potatoes, they hadn't tested french fries. Either way, it would grab attention in Quincy and, they hoped, eventually the rest of the nation.

Lamb Weston, the food plant in Quincy where Patty Martin's husband worked, was the largest frozen-potato processor in the world. J.R. Simplot Company, the other big food company in town, was the nation's biggest exporter of french fries. The northwestern United States supplied 80 percent of the world's fries.

"Lead in Your French Fries?" was thirteen pages, plus attachments. As soon as it was finished, Tom faxed it to Acres USA, "A Voice for Eco-Agriculture." The polemic was, like Tom, smart, direct, and more than a little bit angry.

While I don't have any legal training it seems to me that the state is liable for any damage done to any of its citizens by the actions of the fertilizer companies. The Department of Agriculture is extremely negligent in its enforcement policies, and the people who are responsible for these policies should be held accountable. It is possible that there is enough liability out there to bankrupt state budgets. It looks like the states involved have their asses hanging out a mile on this one.

Witte's paper touched on the Cenex chief scientist.

Max Hammond died on October 24, 1996, he was 51 years old. The cause of death was listed as pulmonary edema. His lungs were destroyed and he was waiting for a lung transplant when he passed away.

Dr. Max was a non-smoking, non-drinking, healthy living Mormon bishop. He was never involved in any occupation that could be considered hazardous to health; yet, he died from severely damaged lungs. Dr. Max firmly believed in what he was doing and he paid the ultimate price for his wrongheadedness.

Shortly after the death of Max Hammond, a team of Cenex attorneys from Minneapolis, Minnesota, came to Quincy, Washington, and gathered up all of the materials that pertained to his work, and took it back to Minnesota.

Later, Max's brother told me the family was enraged by the suggestion that Max had poisoned himself. I asked him how Max got sick.

"We don't know what happened to him; he had problems with his lungs for years and years," he said. But Max always liked working for Cenex. His brother wanted to make that clear: Max didn't blame Cenex.

And Max Hammond had a lot more credibility with the good people of Quincy than Tom Witte did.

Tom concluded the paper with a personal note.

In my own farming operation, it has been a long and difficult process to recover from the losses I suffered. Not only did I have farmland that was unproductive, I also had problems with the dairy cows caused by feed grown on contaminated soil. I personally suffered from numbness on my feet and legs, dizzy spells, and difficulty breathing. Dennis, Russ and Duke have all told me about similar symptoms.

Being able to apply manure from 150 dairy cows has revived the productive capability of my farmland. 1996 was the first year since 1991 that I grew crops that had a normal yield. I have been on a cleansing program designed to rid the body of the toxic elements and my health problems have mostly disappeared. Personally I am quite optimistic about the future.

Tom Witte

Patty had refused to read a draft copy of "Lead in Your French Fries?" She didn't want to be blamed. She asked not to be named in it. She was, anyway.

"There's a lot of supposition in there," Patty said. "Maybe a lot of it's true, but we don't have the proof."

When Patty called me, she sound more factual, less alarmist, than the others, and certainly less motivated by her personal finances. That was the main difference between the mayor and the farmers. She had nothing to gain and everything to lose.

"If this is a good practice of disposing of hazardous waste, then everybody should buy into it," the mayor told me. "The farmers should say okay, yes, this is good, thank you."

Reading a letter to street workers telling them to stand out of the fumes when pouring asphalt, Patty thought about the toxins in asphalt. Of course: highways, byways, driveways, acres of parking lots. Asphalt and concrete were places people could put hazardous wastes. The mixture would dilute the wastes and bind it up in a solid form. But with what long-term effect?

She wondered what was in the cement foundation of her house.

"There's big money in this. Who's making all the money?" she asked.

JANUARY 1997 — *SEATTLE TIMES* NEWSROOM

I started phoning offices that ought to know a few simple facts, but they didn't: the Centers for Disease Control—hadn't studied fertilizer; the U.S. Department of Agriculture—didn't regulate fertilizer; the EPA—sorry.

The EPA referred me to Alan Rubin, a staff scientist who called himself "the king of biosolids." Biosolids is a euphemism for sewage sludge used for plant food, but it is not chemical fertilizer. Rubin had worked fourteen years on so-called beneficial uses of biosolids, arguing that the industry waste in sludge was bound with organic material and safe for plant life.

For this, he'd been vilified by antisludge activists coast to coast. I asked him about heavy metals in inorganic fertilizers.

"I'd love to know more because I'm always getting pounded on biosolids," he said. "Every time I bring up other industrial stuff, they say, 'No, no, everything over there is peachy keen.' I don't think so. There is almost zero federal regulation on fertilizer. We take a lot of pride in saying biosolids are the most studied ingredient EPA has. I have never seen a state or federal limit on heavy metals in fertilizer. You could ask the Fertilizer Institute: How well have the human health impacts of fertilizer been studied? Good luck getting a straight answer. They don't have an Al Rubin there."

The Fertilizer Institute had Ron Phillips, vice president for communications, who assured me the industry was being careful. He told me to check out California for proof.

I phoned the Department of Food and Agriculture in Sacramento. California had formed a task force to study the health risks of heavy metals in fertilizer, a manager told me.

Health risks? That's what I was looking for.

"The task force is not adversarial," he quickly added. "The fertilizer industry itself provided the funding. The industry saw there could be some risk, and they want to know the answers, whether it's good, bad, or otherwise."

I asked for a copy of the notes from their last meeting.

The task force was meeting next week.

Within an hour, a fax came across the telephone line with the first documentary evidence of the toxic risks of heavy metals in the stuff farmers and gardeners use everywhere every day. It was a government report from California, credible and quotable; substantiation. Till I saw this, I didn't know if Patty's claims were any more reliable than UFO sightings. Now I knew the unidentified object was real; in fact, ubiquitous.

The proof was contained in minutes from an obscure group of California bureaucrats and businessmen known as the Heavy Metal Task Force:

Currently, there are no standards for heavy metals in fertilizing materials. . . . The Department is presently pursuing risk assessments of fertilizing materials to identify any unacceptable health or environmental risks. . . . In the interim, portions of the fertilizer industry are subject to the regulatory requirements of the federal and state hazardous-materials law.[1]

Heavy metals. Risk. Health.

The California group had tested four fertilizers and two of them had flunked. The two advertised zinc and iron, minor plant foods. They contained ten times more than the hazardous-waste limits for arsenic, cadmium, copper, chromium, lead, thallium, mercury, and selenium.

Luckily for the manufacturers, they said, toxic metal standards applied to wastes, not products. There were no limits in fertilizer products. Had there been, a lot of the products would have to be disposed of in a hazardous-waste landfill instead of a food-growing operation.

Now I knew I had a story, something new and surprising. Put it this way: There was a limit on the amount of lead in paint, but not in fertilizer. There was a limit on the amount of dioxin in cement, but not in fertilizer. There was a limit on arsenic in industrial slag, but not in fertilizer.

The task force notes showed manufacturers asking state officials to declare the polluted skimmings from metal smelters exempt from standards for hazardous wastes if they, too, were put in fertilizer. It was agreed. They would use the loophole of calling a waste a product. The minutes ended:

All parties agreed that a notice to the fertilizer industry, providing the hazardous material regulatory compliance scheme, isn't needed at this time.

I thought somebody else might be interested in a notice of the hazardous-material scheme: the people who make food, and the people who eat it.

FEBRUARY 1997—SACRAMENTO, CALIFORNIA

The door was open to Room A447 in the Food and Agriculture Building across the street from the gleaming Capitol Dome. Inside the room, men in suits surrounded a conference table. We three outsiders walked in and sat down in empty chairs by the wall.

Tom and Duke had risen at five in the morning to drive across the mountain pass and meet me in Seattle to fly south to Sacramento. Patty had the flu, or she would have come to see the Heavy Metal Task Force, too.

They didn't know we were coming. They didn't know who we were, and they didn't ask. We were bigger than bugs on the wall, but we tried to be just as inconspicuous.

We had some idea who they were from the minutes: John Salmonson, chairman, Monterey Chemical Company, a fertilizer distributor; James Joseph, Bandini Fertilizer Company, Los Angeles; Raymond Maul, Helm Fertilizers; Tim McGahey, Agriform Farm Supply; J. Julian Smith, J.R. Simplot Company; Jay Yost, Unocal Ag Products. We counted fifteen industry people and five state officials.

One of the men was talking about something called fly ash. Fly ash is a waste collected from filters, scrubbers, and other pollution-control equipment in coal-fired power plants. The man, whom I did not know, but who turned out to be the head of the state fertilizer association, complained that Midwest and East Coast industries were cleaning up their smokestack emissions only to lay the waste on agriculture.

Pennsylvania industry alone recycled 4 million tons of coal ash and 2.1 million tons of a similar material called flue dust from smokestacks every year. Railcars packed with fly ash rumbled to the West. They dumped it on California farms, the man said, saving the sixty-five-dollar-a-ton cost of burying it in a landfill.

California growers bought the gray chalky material under names like Lime Plus. They spread it around fruit crops and lawns and gardens.

We were impassive listening to this. I sat on one side of the room, Tom and Duke on the other. The speaker explained that much of the West and Southeast needed lime to raise the alkalinity of the soil, especially in wet areas and heavily farmed areas. Liming improved biological activity, nitrogen fixation, and phosphorus availability in the soil. Liming also reduced, at least temporarily, the mobility of heavy metals.

And California needed a lot of lime. Indeed, there were benefits from the trade. The fly ash was cheap or free. But the man speaking to the Heavy Metal Task Force—Steve Beckley, director of the California Fertilizer Association—said the fly ash provided less benefit to soil than natural limestone and posed more risk to people. The ash contained mercury, molybdenum, selenium, and dioxins, and none of the toxics were disclosed to consumers. Some of the fly ash, Beckley said, "may not be suitable as fertilizing materials at all, due to constituents like heavy metals and other unsafe chemical properties."

Risk. Unsafe.

So it was no longer a naive, small-town mayor and broke farmers spouting off; it was the industry and regulators in the nation's number one agricultural state. It was no longer a question of whether this was happening, but how much.

How much fly ash was applied to California land? Where? On what crops? We learned that nobody kept track.

Steve Wong, California's chief fertilizer regulator, informed the group he had been unable to regulate the toxic metals in fly ash because he had no rules covering a material that made no claims of fertilizing value.

"That's where a lot of this stuff is moving onto the agricultural land, as bulk soil amendments," Wong said.

Tom sat there poker-faced and said "Oh, shit," under his breath.

Our eyes were opened to a new loophole in the crazy semantics of waste disposal. Call contaminated fly ash a soil amendment and you avoid even the loose rules that apply to licensed fertilizers. And when the liming effect ends, as it always does, with time and water, unless you add more and more, then the heavy-metal poisons become active in the soil. It was a time bomb, a Trojan horse, and one feature of heavy metals is that they stay in the topsoil, where plants grow.

Salmonson, president of Monterey Chemical Company, sitting at the head of the table, said California didn't want the fly ash. "It's just like sending nuclear waste to Nevada," he said. "They're trying to shove this into agriculture."

I could have written a good newspaper story that day. But there was no other press within sight, so I had the luxury of waiting, listening, and researching till I knew more.

Steve Wong told the group he had been named to cochair a national group of state regulators studying industrial by-products. He said the Fertilizer Institute had started its own heavy-metals task force because industry was worried California would go off the deep end and drag the rest of the country with it.

"They had a hard time understanding this is an industry-driven process," the fertilizer lobbyist Beckley said.

Canada was already watching fertilizer, Wong told the group. Canada was the only nation in the world to set limits on a range of heavy metals in fertilizer. Canada is the world's second-largest wheat exporter after the United States, and it was acting to protect its exports. The World Health Organization was talking about tighter limits on cadmium in cereal grain. Canadian wheat was accumulating cadmium, one of the most insidious and soluble of heavy metals, from fertilizer. The Canadians were concerned that buyers in Europe, more attuned to toxic chemicals in food than buyers in America, might be alarmed.[2]

So: my first confirmation, from industry, expressing concerns that plants absorb these toxic metals.

Wheat. Bread. Pasta.

Nobody knew.

Tom and Duke introduced themselves to Wong during a break. They handed out copies of Tom's paper. "I really enjoyed giving those guys 'Lead in Your French Fries?' just for the shock effect," Tom said later. "Just to let them know we were onto them. It was no longer a secret."

The members of the task force had no idea what to make of the two farmers in jeans or me with my reporter's notepad. We surprised them. Then they surprised us. Reconvening, Wong announced that the Heavy

Metal Task Force would be going into a private meeting. *Members only.* We were asked to leave.

Tom, Duke, and I walked out of the room, and the door snapped shut behind us.

—

Tom and Duke went to an Acres USA conference in another part of Sacramento, where Tom met a man who told him car makers sent contaminated phosphoric acid to fertilizer makers. The acids were unregulated. Tom talked to a lot of people who had no idea this was going on. He said the organic growers at Acres USA were a lot friendlier than the fertilizer chiefs who'd kicked us out.

I stayed at the State Capitol grounds, pulled out my cellular phone, and called a lawyer in Seattle, who called one in Los Angeles, who called one in Sacramento to argue that the task force was breaking the state open-meetings law. While the lawyers argued, I went back inside the building, opened that door, and walked into the meeting room again.

I told Wong my lawyer said it was an open meeting. I sat down and opened my notebook. After a moment's pause, they resumed their business.

I wondered then why they had claimed it was members-only. It was not. There were ten people in the room, but only two were members of the Heavy Metal Task Force. The others were fertilizer industry staff and consultants and officials. Nobody in the room represented food makers or food eaters.

They were looking over preliminary work from the consultant, Foster Wheeler Environmental Corporation, deemed to be secret information. I listened quietly, thinking *Why are they hiding this?* The discussion seemed guarded, as it so often does when people who thought they were among friends suddenly find there is a reporter in the group.

Foster Wheeler was being paid $161,000 to perform studies to show acceptable amounts of arsenic, cadmium, and lead in commercial inorganic fertilizers. The consultant talked about fertilizer mixers, applicators, farmers, children, and fish as "possible receptors" of the heavy metals in fertilizer.

The group's favorite scientist was a small, elderly, unprepossessing man named John Mortvedt. While he spoke softly, they listened. Mortvedt was, I would later learn, the leading researcher of toxic wastes made into fertilizer in the United States.

Mortvedt had published the only major study to say cadmium did not build up over time when it was applied to soils and crops. European studies had reached a different conclusion.

Most recycled industrial wastes that go to fertilizer, Mortvedt was saying, contain zinc or manganese, minor plant foods used on two major crops in California, almonds and grapes. I made a note to check whether anyone was watching out for hidden chemicals in nuts or wine.

We found out later—after winning an argument with Wong about whether or not the notes of task force meetings were public records (they were)—that the heavy-metals group had operated quietly for five years before the day we outsiders walked in. The task force emerged after California voters adopted Proposition 65, also known as the Safe Drinking Water and Toxic Enforcement Act of 1986. That law required businesses to warn consumers exposed to products which, if used as directed, could cause a significant increase in cancers or reproductive problems. The products wouldn't be banned, but they would have to have warning labels to allow consumers to make informed choices. Information was power.

So far, manufacturers had cleaned up poisonous chemicals from ceramic tableware, foil caps on wine bottles, correction fluids, and fingernail polish. Tavern owners in California had posted signs saying alcohol consumption by pregnant women may cause birth defects. Nobody knew about the quiet little task force on heavy metals in fertilizers.

So quiet, so slow; Tom thought the purpose was to delay instead of to warn. "Their tactic is to obfuscate and delay," Tom repeated often.

People who violated Proposition 65 could be fined twenty-five hundred dollars a day. Nobody was fined. Citizens who brought civil suits under Proposition 65 could keep a 25 percent bounty. The only suit so far was quickly settled by fingernail-polish makers.

A 1992 workshop in Sacramento had been the first to mention health risks in fertilizer. "One representative of industry thought there was an exemption from Proposition 65 on fertilizers. However, this exemption is not applicable when a health hazard exists," the minutes said.

It was then that Wong had asked for volunteers to serve on a task force. He got Mortvedt and five fertilizer company officials with direct interests in the results. At the first meeting, a task force member had claimed some fertilizer materials would at least double in price if they were further refined to reduce the elements that could cause cancer and birth defects. And the minutes said, "The task force committee and industry agreed that the heavy metals issue is of nationwide importance and stated that the nation is waiting for someone to evaluate the situation. There are no standards or limits at this time."

Beckley looked over "Lead in Your French Fries?" I asked what he thought. "It has a lot of facts, but it's one plus two equals six," the fertilizer lobbyist said. In his view, the task force would clean up any problems, and they would be minor. "There could be some loss of product, to be sure," he said, "but maybe those products shouldn't be on the market."

Beckley suggested I talk to somebody closer to my home: Richard Camp Jr. at Bay Zinc Company in Washington State. He said Camp was a national leader in recycling hazardous wastes into fertilizer out of a factory in a town not far from Quincy.

I met up with Tom and Duke at the airport. One thing was about to change. We had just witnessed the last meeting ever held by the Heavy Metal Task Force of the Fertilizer Inspection Advisory Board of the California Department of Food and Agriculture. It was soon disbanded.

WASHINGTON, D.C.

Tom had given Patty a disgusting medicine made from cow's colostrum to help her recover from the flu. In fact, it was not intended for human consumption, but seemed to help. Finally the mayor, who was sick and anxious about flying, felt well enough to go.

In February, Patty and Nancy flew to Washington, D.C., for the three-day First National Research Conference on Children's Environmental Health.

"I'm going for unanswered questions," Patty said. "Nobody would tell me what those chemicals did, and this is my opportunity to find out. I'm thinking of my kids and the junior high school."

They flew in a cloud of secrecy. Patty told Glenn and a few others about her big trip. She told Jess Slusher because he was a minister who would say a prayer for her safe return and a member of the town council she wanted to keep in the loop. But she didn't want the others to know. They might try to hold her back.

When Patty and Nancy got to the conference hotel, they saw poster boards with scientific papers tracking the rise in childhood illness and cancers since the early 1980s.

Ho-oly cow! Look at that. Patty said. *The exact same years they've been dumping more of this waste in fertilizer. And you'd expect the first victims to be the weakest ones, the children.*

Hundreds of people were at the conference, important people. And now EPA administrator Carol Browner was rising to give the keynote address in a crowded ballroom. Browner would open up the floor for questions afterward. Patty and Nancy couldn't believe their luck.

Sitting together at a table shared with doctors and dignitaries, Nancy prodded a nervous Patty to one of the microphones. Patty had credibility as mayor. Nancy wanted her to question Browner.

You gotta go up there now, Nancy told Patty as the speech ended.

I don't want to go, Patty said.

You gotta go!

Patty rose and walked toward the microphone at the front and center of the room. Patty didn't know it, but Nancy walked behind her every step of the way.

Patty stepped up to the mike. She was last in line. Patty felt her heart race. If she got any closer to the mike, she thought everybody in the room would be able to hear her heart pounding as she spoke to the head of the EPA.

I'm the mayor of a small community in Washington State, she said. *I know the EPA is concerned with the inert ingredients in pesticides, but what is the EPA doing about hazardous waste in fertilizer?*

Browner paused. *Well, it all depends on what crops you're talking about—*

Patty's microphone jerked to the right. Nancy had been standing behind her unseen—a foot shorter than Patty in high heels—and now she stepped forward and grabbed the mike.

WHAT CROPS? Nancy said in an outraged tone. *PEAS. CORN. BEANS. POTATOES.*

Patty grabbed the microphone back.

Ever since the EPA delegated this authority to the states in 1978, she said, in a calmer voice, *at least twenty-six states have been putting hazardous waste into fertilizer, and I wanted to know what the EPA was doing about that.*

Browner told Patty she wasn't clear about the question.

Some people in the front row were nodding like they knew. They thought it was biosolids. The EPA was used to getting hammered on biosolids, or sewage sludge spread as fertilizer.

Patty said it was not sludge, not biosolids, not pesticides, not Alar—it was hazardous waste going into the type of ordinary fertilizers used every day by farmers and gardeners. And the government was letting it happen without limits, tests, or disclosure.

Browner said she didn't know what Patty was talking about.

Nancy Witte spoke again.

It's in my brother's tank! she said. *We know it's in the fertilizer*

*because it's in my brother's fertilizer tank. There are these metals—
beryllium, arsenic, cadmium . . .*

Nancy said the government was poisoning people by allowing contaminants in fertilizer. Nancy, like Tom, didn't mince words.

*Farmers have unknowingly become the agents in this deplorable
situation, and they and their families are suffering from the highest
rates of severe illness and disabilities.*

It was as if they were shooting bullets at Browner. The administrator turned away, dodging. Browner said she didn't know anything
about it, but she would get someone to find out: Dr. Lynn Goldman,
EPA assistant administrator in the Office of Prevention, Pesticides and
Toxic Substances. Then Browner turned to the other side of the room
and left them standing at the microphone.

(Hugh Kauffman, the famed EPA whistle-blower, told me later that
he thought Browner was lying when she said she didn't know about
hazardous waste in fertilizer. He said everybody in the EPA knew about
the loophole in the law. Kauffman said the traditional environmental groups "wheeled and dealed" to give industry a loophole for
land disposal of hazardous wastes in return for industry support of
recycling.)

Patty compared their comments to dropping a bombshell. People
surrounded Patty and Nancy during a break. They'd known about pesticides and sewage sludge but not this, not fertilizer.

Patty was astonished by one reaction most of all. That was the
reaction of belief.

"Nobody was in disbelief!" she exclaimed. "People had questions—
'What proof do you have?'—but nobody disbelieved us. They said EPA
hasn't been giving information on the inert ingredients in pesticides,
either."

So what proof of harm did they have?

"I think that's the million-dollar question," Mayor Martin said.
"The government is banking that even if it *is* causing something, you're
never going to be able to *prove* it, instead of erring on the side of caution and getting the science and the facts together before you do it."

Patty and Nancy looked around. They saw Dr. Goldman, the EPA deputy, walking out the door across the room. They thought Goldman was trying to avoid them. They took off across the room but lost sight of Goldman.

Patty went to the top of a set of escalators while Nancy patrolled below.

Patty, there she is! Nancy hollered, pointing at the escalator Goldman was riding up.

By then Patty had already started riding down the other side of the escalator. Patty Martin and Lynn Goldman crisscrossed, almost close enough to touch but going opposite directions on the moving stairs. Nancy Witte got on the up escalator behind Goldman; Patty turned and started bounding against the flow.

The mayor raced up the down escalator as fast as she could in her high heels and dress, laughing all the way to the top. Patty and Nancy both caught up with Goldman in the hotel gift shop on the second level. They asked her when she'd talk with them. Goldman, surprised, for she was just taking a moment to buy a gift, told the Quincy women somebody would phone them soon. She promised.

Okay, Patty said. *We'll hold you to that.*

From that time, at every break in the conference, every lunch, every dinner, every time Patty or Nancy was standing in a hallway waiting for a bathroom, people were talking with them, intrigued with what they had said. The difficulties before and after were bathed and washed and forgotten in the glow of the moment. The two women from Quincy met the Sierra Club and the Natural Resources Defense Council, the group that stopped Alar.

Donald Paglia of the UCLA School of Medicine Department of Pathology and Laboratory Medicine offered the services of his laboratory to investigate heavy-metal intoxications. Paglia was working with a National Institutes of Health grant on the effects of low levels of lead on children. He had no idea about lead in fertilizer.

"I certainly never heard of anything like that before, or since," he said later. "I had the feeling their comments were taken with a certain degree of skepticism by all the officialdom there, and there was a lot of Washington there. Nobody seemed to know anything about it. But I talked to Patty and to Nancy, and I was curious, because as wild as the tales we hear today, some have an element of truth in them.

"It's something I never encountered in any of my reading, either scientific or open. It's fascinating. I hadn't known that. I had the sensation that this is quite plausible, and it's very quiet, and *that* worries me."

Somebody from the W. Alton Jones Foundation said he'd come out to Quincy and investigate. Somebody from *Consumer Reports* magazine said it could test fertilizers. A doctor with the council of state epidemiologists said she'd help.

Patty and Nancy flew home on a cloud of euphoria.

But in the end, like all clouds, it dissipated. The people who promised help were busy with other projects, like pesticides. *Consumer Reports* got a twenty-five-million-dollar grant to test food, but not fertilizer. Only the EPA took action—and not the sort of action Patty or Nancy wanted.

SEATTLE TIMES NEWSROOM

While Patty and Nancy were in the nation's capital, I was on the World Wide Web. I finally knew, after Sacramento, what questions to ask. I put some words in an Internet search: "hazardous" "waste" "fertilizer."

And in five seconds, I found a tattletale site on the Web. Somebody inside the fertilizer industry was blowing the whistle on the rest of them.

Cozinco Sales was a small fish in the fertilizer ocean, but it snapped at the bigger ones recycling industrial wastes.[3] The first screen of the Cozinco Web site said:

Ours vs. Theirs
The difference is black & white

Pointing my mouse at a picture of black-and-white granules on the screen, I clicked.

Environmental Risks
Get the Lead Out!

Agriculture is being used by some industries as a dumping ground for their hazardous waste—lead, cadmium, other heavy metals.

Click.

Sources of these by-products are flue dust and baghouse dusts from zinc smelters, copper recycling plants, steel mills, and other industries. Another major source is spent acids from the galvanizing industries.

Click.

Disposal of hazardous waste. . . .
At what cost to farmers?

Click.

If the granular zinc product is white, you don't need to worry. If it's not white (gray, black, brown), you must have it tested for solubility and heavy metals. . . . Because there are no labeling requirements for zinc fertilizer, you have no idea what amount of lead and cadmium could be in a product. Look at this Product Comparison table. *We'd be glad to do a* free test *for you.*

A fictitious farmer on the Web site, named Mort, said the high-lead, low-soluble products were lucrative waste-disposal schemes but poor fertilizers. *Click.*

Mort Says about Fertilizer vs. Shotgun Shells
"I've been duck hunting in a hundred-acre field by the house since I was a kid. Several years ago they took the lead out of my shotgun shells. Said it was killing the ducks. When I found out the 50 pounds an acre of zinc fertilizer I was putting out contained 3 percent lead, I did some figuring. I would have to shoot 2,500 shells over that hundred-acre field to put out the same amount of lead that I'm putting out with my fertilizer. I wonder what that could be doing to the ducks."

I clicked the link to e-mail the company.

Kipp Smallwood, sales manager for Cozinco, emerged from cyberspace. He'd set up the Web site two weeks earlier, and he told me he got a quick response from the executive director of the Far West Fertilizer and Agrichemical Association. "Stop pouring gasoline on the fire," the man had told him. Smallwood's voice was animated on the phone as he talked about the confrontation with the fertilizer association that he *thought* was supposed to be working for him and for the farmers.

"I told him there are things going on that are bogus and I won't be quiet because I think this is unsafe." His voice went higher. "It goes from Far West to the very top in the United States fertilizer industry, to the Fertilizer Institute in Washington, D.C."

His voice slowed. "I'm an ankle biter. But I'd be painted with the same brush if I didn't stand up and be counted."

Kipp Smallwood grew up in the business. His father, Don, also worked at Cozinco, based in Denver, operating three plants making zinc to blend with fertilizer.

"People try to find the cheapest source of products they can find,"

Kipp Smallwood said. "Urea is urea. Nitrogen is nitrogen. There aren't any heavy metals in those products. When you get into micronutrients, that's where it gets ugly. Zinc, iron, manganese, copper, boron, magnesium—all of 'em, or almost all of 'em, come from recycled industrial waste.

"All this stuff is black, brown, gray, all these recycled products that don't dissolve in water because they're not soluble and they haven't gotten rid of the toxic chemicals."

Smallwood's company paid for high-quality raw material, paid to purify it further, and paid to dispose of the industrial poisons. Cozinco reduced the lead content to less than fifty parts per million, clean enough for animal feed, and charged three hundred dollars a ton for zinc.

The sham recyclers of the world cut corners, he said. Their zinc had thirty thousand parts per million lead. *Three percent!* Nobody had *proved* it was dangerous, but it was. That's why they could charge two hundred dollars a ton and still make a good profit.

Smallwood did the math: The price difference between pure and dirty was one hundred dollars a ton. A ton of zinc covered two hundred to four hundred acres. The difference was twenty-five cents to fifty cents an acre. Fifty cents an acre on five thousand acres of potatoes is twenty-five hundred dollars.

"That might be enough to swing a bid. But the farmers never see the savings, the dealer does. Or the farmer might save a few cents an acre but they'll get junk and nobody will tell 'em. The lure of the dark side, you know, the dollar. They get away with it because there's nobody watching, nobody testing. I think there's stuff you couldn't imagine going on the ground."

Kipp spoke thoughtfully and composed each sentence.

"We don't have many friends," he said. "It's hard for people. We kind of fight a lonely battle out here."

In recent years, after the J.R. Simplot Company started checking its supplies for toxic metals, Cozinco won more sales. Simplot found dangerous levels of lead in the air of a fertilizer-blending facility in Idaho. The company would have had to give workers respirators and regular blood tests if it didn't insist on cleaner material.

Smallwood knew about the California Heavy Metal Task Force. "That's a joke," he said at the end of our phone call. "I don't even talk to them. That's the industry group. What they think is all right, I don't.

"I've gone to meetings, done all that, and got crushed by my competitors. Now I'm using the Web site to go straight to the farmers. They'll understand."

CHAPTER 6

THE MAGIC SILO

"WHEN IT GOES INTO OUR SILO, it's a hazardous waste. When it comes out of the silo, it's no longer regulated. The exact same material. Don't ask me why. That's the wisdom of the EPA."

Richard Camp Jr. talking. One of the top hazardous-waste-to-fertilizer dealers in the nation, Camp knew why. He knew it better than anyone: Why EPA stopped regulating that waste partway down the silo. Why there was so much money to make.

Camp exuded schmooze. As he had sold steel mills on the idea of paying him to take their worst waste, and sold farmers on paying him for the privilege of putting it on their land, Camp, personally, had also sold the EPA on the wisdom of the idea. No one found out till later. Dick Camp was that good a salesman.

It was a spring day in 1997 when Camp visited me at the *Seattle Times*. I'd called him after hearing about him in California. He wanted to sell me on the idea that there was no story here. Camp strolled into the newsroom talking enthusiastically, greeting the managing editor, Alex MacLeod, who'd been a fraternity brother of his. Later Alex told me he wouldn't be surprised at anything Camp was up to, and rolled his eyes.

My desk was piled with files on the investigation that threatened Camp's operation. He kept glancing at the files as he talked with me. I didn't mind the free look. I've always wanted to be as open as I can with investigative targets because I want them to be open with me. We sat down in an editor's glass cage near Alex's. I opened my notebook, asked one question—basically, "What's going on?"—and Camp talked for the next hour.

Dick Camp was big—six foot three, 250 pounds—and voluble. His father had been even bigger and friendlier. His father's story was the start of the hazardous-waste-to-fertilizer business in America.

Richard Camp Sr. had grown up poor in Seattle, worked through college, and learned the trade of boilermaker. While he was cutting steel, he was throwing off ideas like sparks. He bought a little land near Commencement Bay in Tacoma and used it as a parking lot. When the war started, the Tacoma shipyard started to build a lot of flattops. Camp had the closest parking lot. He sold it to the government and used the profits to open a galvanizing shop that won navy contracts.

Camp and his workers rustproofed chains and anchors. They dipped the steel parts in a vat of molten zinc, and as part of their work, from time to time, they skimmed off the top of the vat a layer of impurities left by the steel in an oxidized scum. The zinc skimmings had no further purpose. They filled fifty-five-gallon barrels out back in the yard behind the shop.

Camp made barrels of money, then the war ended and so did the government contracts. He left the barrels of zinc skimmings in the yard.

Camp tried other work: plating chrome, bathroom fixtures. He almost went bankrupt. He went back to boilermaking in shipyards around Puget Sound. One day, his son told me, Camp happened to see some people pour a white powder out of a bag into a fifty-five-gallon drum and fill it with water. Curious, he struck up a conversation and learned it was zinc sulfate—fertilizer for apple trees. And apples were the state's leading crop.

He got a new idea for the barrels of waste behind his shop in Tacoma.

A new industry was born.

With help from a friend who knew a little chemistry, Camp started to make the zinc skimmings into zinc sulfate. He didn't think about "recycling" or the "environment" or "dioxins." It was 1954. Nobody did.

Camp Sr. just thought he had some wasted stuff sitting around in the barrels that he could use as a raw material to make some money. He sold an alternative to the white powder of virgin zinc. Camp started selling the processed waste to apple growers.

"The fruit tree industry has always been very progressive, a little ahead of the row-crop guy nutritionally and chemically," Dick Camp Jr. explained. "There was always a lot of money in apples. So Dad started selling zinc sulfate to the orchard guys in Wenatchee and Yakima." When his own barrels were empty, he bought zinc skimmings from the galvanizing companies in the Seattle area.

"Then there was another guy up here someplace that was making it, so Dad said 'Wait a minute, we'll beat the freight and make it in Yakima and we'll have a leg up on 'em. Save a penny or two a gallon on freight.' He moved to Yakima in the early sixties."

Turned out the "other guy" was named Garnt Niewenhuis, and he was a chemist who owned Western Processing Company south of Seattle. The company mixed hazardous waste from various industries and spent nitric acid from Boeing Company to make a fertilizer in the 1960s and 1970s. More than two hundred businesses got rid of their waste oils, acids, flue dust, pesticides, solvents, and zinc dross that way. Niewenhuis worked for years in obscurity making fertilizer before federal agents raided his plant for pollution violations. Niewenhuis was prosecuted under federal hazardous-waste laws. His fertilizer company became the most notorious Superfund site in the state.

But Camp didn't tell me that.

No, Dick Camp Jr. was proud of his family business. It was innovative, legal, and fruitful in more ways than one.

His father expanded the market. He developed a dry zinc product to mix with regular fertilizer for row crops. He educated the farmers on zinc deficiencies. They started asking their fertilizer blenders for more zinc. And when there weren't enough zinc skimmings around the Pacific Northwest to fill the demand, Camp Sr. came across a new source of zinc.

Flue dust, filtered from smokestacks at steel recycling companies, was lower in zinc and higher in heavy-metal contaminants than the zinc skimmings that Camp had been selling. Camp discovered he could be paid by the steel companies to take their flue dust away. They paid him as much to take it away as the farmers would pay him to put it down.

"I believe he was the first guy to use it," Dick Camp Jr. said proudly.

Dick Camp Sr. made money coming and going. He had tapped into the biggest recycling story in history, at least the biggest since manure: secondary steel. Like much of America, it was built around the automobile.

Today in the United States, 60 percent of junk cars are recycled in electric arc furnaces.[1] Every ounce of steel west of the Rocky Mountains is used, secondhand, recycled material. None of it comes from a mine. The steel industry takes forty-five million tons of ferrous scrap off the American landscape each year.

Every time the steel is melted to purify its main ingredient—iron—the contaminants are concentrated in a toxic waste. The zinc, cadmium, chromium, arsenic, lead, and dioxins go to fume at a lower temperature than iron melts. Up the chimney they rise, every impurity unwanted by steel recyclers. To prevent air pollution, the electric arc furnace dust is cooled and collected in a structure called a baghouse.

The dust is typically 10 to 20 percent zinc from galvanized parts of the car, like the door handles. It's about 3 percent lead, 0.5 percent cadmium, and sky-high in dioxins from incinerated plastic. In the end, about 1.5 percent of the weight of the recycled steel ends up in the bag-house. The forty-five million tons of scrap metal leaves 650,000 tons of hazardous waste a year.

Dick Camp Sr. tapped into what had been the worst garbage at the end of the process. He advertised the zinc in baghouse dust. And he did very well. He was quite the salesman, like his son, whose story cast a vision of entrepreneurial success.

Richard Camp Jr. had graduated high school in 1965 and joined the Army Reserve in college to avoid the draft. Camp hadn't been antiwar; he'd thought President Johnson was too soft and he didn't want any part of that. He got a degree in economics from Whitman College in Walla Walla and married a zoology major and violinist named Marilyn. Then, when his father died unexpectedly, he took over the recycling business. By then it had moved operations from Tacoma to Yakima to be closer to the farms and orchards of the Columbia River. The company changed its name from Bay Chemical to Bay Zinc. (There is no bay in Yakima.) Over time the Yakima site became contaminated.

Then the company moved to a hop field served by a sewer line and a rail spur outside Moxee City, population eight hundred, eight miles from Yakima. Camp Jr. had the touch. He developed markets throughout the United States and the world.

Then, in 1976, President Gerald Ford signed the first law to try to control hazardous waste dumping, called the Resource Conservation and Recovery Act.[2] The law required the EPA to oversee the treatment, storage, transportation, and disposal of hazardous wastes, "cradle to grave." It set in motion a twentyfold price increase in fees at hazardous-waste landfills over little more than a decade, increasing pressure to recycle. The EPA mantra was "reduce, reuse, recycle." The law was so complex that the government hired a whole staff of workers to do nothing but answer questions on how to maneuver through it.

Camp became an expert on the new law. He knew it better than the lawyers. He made it his business.

Of the many ways to recycle waste, the EPA was most alert to disposal on the ground. The agency watched for companies trying to use recycling for surrogate dumping. The EPA said they had to make a viable product, stabilized and safe. Otherwise it was sham recycling. That was rarely enforced.

In 1978, the EPA briefly focused on recycled waste in soil conditioners.[3] If a hazardous waste was made into a product for the soil, the agency proposed, it should have strict limits on the amount of arsenic, cadmium, and lead. A distributor could exceed the limits only if buyers were warned with labels saying the product was made from hazardous waste, listing the levels of toxics, and advising users not to apply it to food crops. This proposed rule died a quiet, early death.

By 1988, when the EPA finally wrote a complete set of rules for the hazardous-waste law, Ronald Reagan was president. Vice President George Bush chaired a panel on regulatory relief. EPA administrator Anne Gorsuch eliminated the hazardous-waste enforcement office. By then Bay Zinc had been doing business for almost a quarter of a century.

The American rules were filled with loopholes and anomalies.[4] Camp and his colleagues in the growing micronutrient business exploited them wherever possible. They even got the EPA to add a new loophole: electric arc furnace dust K061 (named after its hazardous-waste designation) would—despite hexavalent chromium, lead, and cadmium—simply not be considered a hazardous waste if it was used to make fertilizer.

The K061 exemption required a chemical change to the material, but I would later learn the change was cosmetic. Bay Zinc simply mixed sulfuric acid and water with the hazardous waste to turn a powdery dust into solid granules. Camp said this followed the federal law. But obviously in the end, when the granules dissolved in the soil, they released all the toxic metals that had been collected in the steel-mill chimneys. After all, the purpose of fertilizer is to release chemicals in the ground.

"Our material has a specific exemption," Camp said. "The EPA likes it. I don't see why that's news."

I did. It was news because people didn't know about it.

Camp said he always wanted to "take the bad things out" of the steel-mill flue dust, but he and his father could never figure out an affordable way to do that. They relied on dilution with the land to render the toxic chemicals harmless.

Camp told me to put the contaminants in perspective. "We throw away far too much stuff in the U.S. We make a big deal of recycling newspapers and beer cans, but when it gets into real serious recycling, people get worried because it has a little bit of this and a little bit of that," he said.

I could imagine Camp making the same pitch to the EPA. He had moved up the ranks of the American fertilizer industry and won election to the board of directors of the Fertilizer Institute. When contaminants in fertilizer became an issue briefly after Greenpeace exposed an illegal shipment to Bangladesh, the board turned to Dick Camp Jr. for advice on what to do. He had the weight of the agrochemical industry behind him, and he used it.

I learned later that Camp had been the key person lobbying the EPA for the fertilizer loophole in 1988. The only tests the EPA got were the ones provided by Dick Camp. He knew exactly why the EPA did what it did.

John Mortvedt was another part of the reason why the same stuff could be a hazardous waste at the top of the silo and a fertilizer at the bottom. Mortvedt, then a Tennessee Valley Authority scientist, had sent the EPA a copy of his 1985 study on wheat and chard absorbing heavy metals at levels he did not think were significant. There were no other U.S. studies. There is no evidence that the EPA looked at studies in Europe and Australia showing significant concern with cadmium transferring from fertilizer to food.

And now the American hazardous-waste products were being sold for farm use around the globe.

The American steel industry had been locked in a battle with Korea in the late eighties. The loophole saved Big Steel a little money. While it might cost two hundred dollars a ton to dispose of its flue dust in a regulated landfill, it would cost half that amount or less to give it to a fertilizer maker like Camp. With 650,000 tons a year of hazardous waste, the potential savings added up to tens of millions. At the very least, the competition from fertilizer makers held down the charges imposed by hazardous-waste landfills.

From Camp's perspective, a ton of hazardous waste from a steel mill represented seventy-five or one hundred dollars in Bay Zinc's bank account. A trainload—four hundred tons—was thirty thousand or forty thousand dollars. By adding acid and water, he could roll the ash into granules and sell it as fertilizer for one hundred to two hundred dollars a ton. The industry grew from the EPA exemption.

At the steel mill, it was hazardous waste.[5] In the railcar, it was hazardous waste. Going into the top of the silo at Bay Zinc, it was hazardous waste. All subject to "cradle to grave" monitoring and treatment that the EPA set up for hazardous wastes.

Then it changed. Camp knew why.

I came to think of it as the magic silo. When it came out the bottom of the silo, it was no longer considered a hazardous waste, but a fertilizer material.

The same stuff.

I asked Camp if he would show me around Bay Zinc Company.

Can't, he said. *Trade secrets.*

Yeah, I thought. *Right.*

Camp never let me see the inside of his factory. I think he didn't want anybody to see. Because Bay Zinc was not just a maker of fertilizers. It was also a federally licensed handler of hazardous industrial wastes.

Blu Min Zinc fertilizer was a mixture of water, sulfuric acid, and steel-mill flue-dust hazardous waste. The workers added water and acid to the gray ash in a steel drum they called Big Bertha. The drum rotated like a cement mixer with a rumbling seventy-five-horsepower engine rolling the thick solution into rounds of fertilizer. As the material dried, it passed over a screen to remove odd-size granules. Sometimes a high-zinc, low-contaminant industrial waste like brass dust, which Camp had to actually pay for, was added to the mix to raise the zinc content to the guaranteed 18 percent. Workers put a light coat of oil on the granules to reduce the dust and make them shine and easier to sell.

The gray material was poured into fifty-pound bags and one-ton bags. Blu Min Zinc was sold wholesale to giant fertilizer companies.

"It's like selling washers to Ford Motor Company," Camp said.

The blenders mixed the zinc with other materials for general farm use. The black and dark gray granules you might find mixed with white, blue, and other colors in a bag of fertilizer may be Dick Camp's contribution.

By now, Bay Zinc and seven other companies in the United States made over 120 million pounds of hazardous waste into fertilizer each year. It filled about half the market for zinc in American agriculture, a market steadily expanded by salesmen like the Camps.

Patty and I would later learn that Bay Zinc alone, between 1990 and 1996, had disposed of one and a half million pounds of lead, eighty-six thousand pounds of chromium, and nineteen thousand pounds of nickel in its fertilizer products.

After Dick Camp left my office, I phoned Kipp Smallwood of Cozinco to ask him about what Camp had said.

"The whole thing kills me, that they can wave a magic wand over it, and it's no longer a hazardous waste, it's fertilizer," Smallwood said. "EPA's never given it another glance. 'Reuse, recycle,' that's all they see. Meanwhile the whole fertilizer industry has no labeling. Who's watching the heavy metals? No one."

Smallwood was familiar with everybody in the industry.

The biggest haz-waste recycler was Frit Industries, based in Alabama, with operations in Nebraska, Iowa, Arkansas, Virginia, and Brazil.[6] While Bay Zinc had one little plant, Frit was everywhere. While Bay Zinc sold two or three zinc products, Frit sold scads of zinc, manganese, boron, copper, and iron to fertilizer mixers. It had some of the most contaminated products ever tested.

Smallwood had a nice way of putting things. "Frit Industries," he said, "makes Bay Zinc seem like Boy Scouts doing the right thing."

—

Before I could call Mayor Martin to tell her about my tutorial, Craig Smith, vice president of the Northwest Food Processors Association, called me.

He asked about the Quincy mayor. He'd heard I was working with her. The questions Martin was asking were, as he put it, "of concern."

Was it true, Smith wanted to know, that hazardous wastes were being put in fertilizer?

I told him the short answer was yes. For the long answer, ask Dick Camp.

Then I asked Smith to comment on this. He said he would comment after he learned more about it.

When Smith hung up, he thought, *Alar*.

Since Alar, the food and chemical industries had learned to try to stop food scares before they started.

MARCH 1997—QUINCY, WASHINGTON

Two state officials came to town to meet with Patty Martin and the farmers.

We're getting waste dumped on us, and we have no idea what it is, Dennis DeYoung told them. *Stand up and say this is not acceptable to agriculture.*

But the officials told them it was acceptable.

What is acceptable to you and us is different, Tom replied. *You don't even know what's in it.*

Dennis asked, *What can you not put in fertilizer?*

The Washington officials could not answer. They did not know. The answer was nothing.

Dennis said, "Their whole purpose in this isn't to enhance farming. It's to dispose of unwanted materials. What they're doing is moving the liability from industry to the American farmer. And that is not right. I feel the farmers have the right to know what goes on their ground because it's still their ground."

Or in his case, it used to be.

Dennis hadn't heard anything from the state court of appeals and wasn't counting on any help to get his hundred acres back. His ground was gone.

A few days later, Tom Witte called. He'd been reading, thinking.

"I did figure this out," he drawled. "This whole thing with fertilizer: It's the acid. Acidifying whatever to make fertilizer. Using the spent acid is the dangerous practice. If you put this on the ground in an acid form, a lot more of it is instantly available to the plant. Industries use sulfuric acid and phosphoric acid, then it goes back into making the fertilizer.

"Organic materials tie up the heavy metals, basically. Tie them up until the organics turn into humus. They will never be in as active a form as they will when they're treated with acid. It's something any chemist should understand."

I checked it out. Tom was correct. Composts, manures, mulches, and organic fertilizers release their nutrients slowly and increase the biological life of soil. Acidic synthetic fertilizers give a quick hit.

Patty Martin kept phoning and faxing me with tips, leads, hunches, suspicions, and wild guesses. Clark Kent himself couldn't keep up with all this. One day she sent a newspaper clipping about wastes on the Hanford Nuclear Reservation. Patty circled a paragraph midway through the article.

Hanford officials also cleverly managed to unload 183,000 gallons of contaminated nitric acid on a British firm, which saved U.S. taxpayers another $37 million in anticipated treatment costs.

Nitric acid, she thought: nitrogen in fertilizer.

Soon after that, Patty met with Gary Chandler at the Quincy Town Hall. Chandler was a state legislator and owned an apple orchard in Moses Lake, an hour's drive from Quincy. He was a Republican, but Patty never thought this was partisan. She gave Chandler the article. She told him about her suspicions.

He listened, wondering, *What the hell is she talking about?* Chandler had never heard about hazardous wastes being used in fertilizer. He was surprised. He said he'd look into it and get back to her.

If there was a problem, Gary Chandler could fix it. He was chairman of the House Agriculture and Ecology Committee in the legislature.

OLYMPIA, WASHINGTON

Patty had woken at four in the morning. She never slept well before traveling. Leaving Glenn asleep, she'd slipped out of bed, gulped down orange juice and coffee, and left home again. Glenn would make sure the children had a good breakfast before school. She would call him on the cellular phone to make sure.

The sky was asphalt black. She passed the food and chemical companies to her left. Leaving Quincy, she crossed the Columbia River at Vantage Bridge and the Cascade mountains at Snoqualmie Pass. The sky was cement gray when she arrived at the state capitol in Olympia before eight o'clock. Men and women in wool overcoats walked among the marble buildings.

The House Agriculture Committee was taking testimony on a new law sought by the wood-products industry to use the waste from newsprint recycling as a "soil amendment" on farmlands. This would be Patty's coming-out testimony. Her chest felt tight, and she was tired.

She'd been told about Senate Bill 5701 by Duke, who'd been tipped by somebody he met in Olympia. They'd learned Chandler, the powerful legislator and orchardist, was supporting the bill. Another hope for help, gone.

So Patty had phoned Glenn's boss at Lamb Weston, white-haired Dwight Gottschalk, to tell them there was a proposal to legalize putting industrial waste in fertilizer.

Lead, cadmium, chromium, mercury, she'd told him.

Patty respected Dwight's judgment. Dwight knew food had to be kept clean as a matter of the precious, fragile trust by the buying public. And he still treated Patty as a friend. Not long ago, Dwight had helped Patty get a twenty-five-thousand-dollar grant from ConAgra for a new town swimming pool. They'd been pictured in the newspaper shaking hands as Dwight gave Patty the check.

So Patty felt safe telling Dwight this was important: *The food products industry will be hurt by industrial waste on farmland. The heavy industries get rid of their mess, but it goes on food.*

Dwight was glad she'd called. He told Patty to call Craig Smith, vice president and lobbyist for the Northwest Association of Food Processors, and repeat the warning. She did. Smith said he'd look into it. He called Scott McKinnie, executive director of the Far West Fertilizer and Agrichemical Association in Spokane. Then he called Patty back.

Don't worry, Smith told the mayor. The fertilizer group knew about the bill. Companies were recycling paper wastes, that was all. They were careful. It was safe.

Patty flushed, remembering. She'd felt Smith stabbed her in the back. Another hope for help, gone.

Tom Witte had worked against the bill, too, using his own best weapon: the written word. The *Wenatchee World* had published Witte's letter:

If the idea of toxic waste being disposed of through fertilizer which is then applied to farm ground allowing for uptake of heavy metals into plants and eventually into people, is such a good idea, why doesn't everyone know about the practice? Why won't any of the bureaucrats representing any of the government agencies speak to us openly about the practice? They are trying to pass it off as a wood ash bill when it would actually allow anything to be used in fertilizer. . . . If this bill was to pass it would be the biggest health risk since Hiroshima.

Witte was not fond of understatement.

Patty had called an environmental activist in Seattle, Laurie Valeriano, policy director of a group called the Washington Toxics Coalition. Patty told her she needed to jump in and fight this, stop it, let the people know. Valeriano replied that she didn't know anything about Senate Bill 5701 and could not help.

You don't want me there. They don't like me.

So when Patty walked into the legislative building shortly before eight o'clock, she thought she was the only person who was going to try to stop the bill. And she was.

The hearing room was crowded with people sitting in folding chairs—industry, she thought to herself—but half the senators' seats were unoccupied on the raised platform at the front of the room. There was a friendly face: Jim Justin from the Association of Washington Cities, who had put her in touch with me. Patty told Justin she was worried about testifying because she didn't have the blessing of her council.

Relax, Justin told her, just say this is your own opinion and you're not representing the city. It was advice Patty forgot as soon as she was called to the table at the front of the room to testify.

"I am Patricia Martin. I am the mayor of the City of Quincy. And I'm here on behalf of the citizens of my community and for members of the agricultural community who may not be aware of this legislation.

"This legislation has a broader impact than simply allowing wood by-products to go into commercial fertilizers. It's garbage, it is recycled material, it is industrial waste, it is ashes and slag, and we're talking about now being able to put all these into commercial fertilizer because they can be called a fertilizer and they're no longer looked at as dumping solid waste on the land.

"The only chemicals that farmers are aware that they're getting in their product are nitrogen, potassium, and phosphorus. It doesn't identify what those other chemicals are. And I might remind you that this is not like biosolids. Biosolids go to a farmer, the farmer is aware that he is receiving them, a farmer is *paid* to receive them, and biosolids are heavily monitored. You know exactly what you're getting in the way of heavy metals or whatever solid waste may be within those. This provision to make solid waste legal to apply to land in commercial fertilizer, we're now talking about food crops. And we're talking heavy metals.

"That bag of soil amendment passed around contains heavy metals. And we're talking about things that bioaccumulate in the soil, bioaccumulate in food, and eventually bioaccumulate in humans—and children in particular are at greatest risk."

Patty continued, "Adulterating fertilizer without advising the farmer what the ingredients are that he is getting is legalized fraud. I mean, farmers have a right to know what it is that they are applying to their land."

Jim Pendowski of the Washington State Department of Ecology spoke next.

"From the department's perspective," he said, "we are increasingly seeing materials that are being derived from the waste stream that are being reported to have commodity or product value. We find people purporting waste to be product and then using that to avoid regulatory requirements under the solid-waste law. So I think Mayor Martin's concerns are well founded."

He promised that the department would make sure the recycled materials were safe to use as fertilizer.

Afterward, Gary Chandler walked by and said, *Don't worry. We've got it all taken care of.* Patty didn't know what that meant. She didn't trust Chandler.

Patty turned down an invitation for lunch. She strode to her car and drove straight back to Quincy, pausing at a McDonald's drive-through for lunch, arriving back home by four o'clock in the afternoon, just after the four children got home from school.

APRIL 1997 — *SEATTLE TIMES* NEWSROOM

The glamour of investigative reporting: I worked on the telephone at a cluttered desk in a windowless corner of the newsroom to check out the outlandish theories of the Quincy water group. I was fascinated with these unsophisticated, distrustful, and zealous people.

Over time, I became convinced they were onto something new. What, I could not say. My task was sorting the kernels of fact from the fields of accusation.

Duke showed me the first proof of pervasiveness in a package that arrived one day from the department of ecology bureaucrat in Spokane. It was a thousand pages of material on hazardous-waste storage, transport, and recycling, all public record.

One brochure showed Ma and Pa Kettle on the cover holding plastic bottles and newspapers for recycling; the information inside was aimed at heavy industry with troublesome toxics. The federal government had set up industrial-materials exchanges after World War II to introduce waste generators to people who would take the waste, "like a blind-dating service," Duke said.

The industrial-materials exchange, or IMEX, provided free listings and confidential referrals. The government was the matchmaker. Twenty-six states had set up industrial-materials exchanges. Washington's IMEX was sponsored by the health department and the ecology department: The same people who were supposed to watchdog the wastes were helping move them around, confidentially.

Duke had found the fundamental loophole that let toxic waste be added to fertilizer.

He read aloud: "Under state and federal law, wastes that can be used as effective substitutes for commercial chemical products do not have to be manifested or handled as hazardous wastes."

That was difficult language for a simple idea: If some of the chemicals in a toxic waste can be used as a plant food, the toxic chemicals go along for the ride.

"They just call a hazardous waste a fertilizer," Duke said, "and by some kind of magic it is no longer a hazardous waste."

There was evidence, too, that a cottage industry had grown among the private brokers who got in the middle of industrial-waste exchanges. Duke called them pimps. In one case, Burlington Northern Railroad bragged it had saved over ten thousand dollars in testing and disposal costs for twelve hundred gallons of used sulfuric acid and thirty-four hundred gallons of caustic soda by giving it to a trucking company. No one knew what the trucking company did with it.

The farmer Tom Witte best described the practice in understandable terms. He was patient as I challenged his theories.

"What's crazy about the department of ecology is they are so much on the side of the companies," Tom drawled on the phone one day. "It's a way to get rid of it. They figure if they spread it around on the ground enough, it'll be okay. It is okay if it's not harmful. But they're never checked for harmful substances."

Was this a Quincy problem? A state problem? "No, no, no, no," old Tom spoke softly and punctuated his sentences with low chuckling as if it was all so absurd. "This is part of the system. This is how they get rid of the stuff. This is how they get it done. They put it on the ground. I think when you dig deep enough, you find out that everybody is involved. EPA is involved. Everybody is involved."

I checked.

Witte was right.

Environmental agencies were promoting toxic-waste recycling, and fertilizer companies were helping to spread it around and mix it in. They were not checking for toxic chemicals. They could not know what was safe. There was almost no scientific research on the subject. There was no way for buyers to know what they were really getting.

JUNE 1997 — QUINCY, WASHINGTON

From home, Patty phoned and wrote people she had met at the Research Conference on Children's Environmental Health. *Help!* She mailed off copies of the test results showing lead and mercury in soil, fertilizer, potatoes, hay, and beans from Quincy.

"I believe strongly, that unless we stop hazardous waste from going into fertilizer, the only standard that will govern this practice will be greed," she wrote.

Patty called me almost every day. She made outrageous claims about some people that I could never repeat in print. I encouraged her to share her wildest suspicions and assured her I wouldn't print them unless I could prove they were true. Otherwise, my newspaper would be sued for libel even before she'd be sued for slander.

Another journalism aphorism holds that 90 percent of interviewing is the ability to ask just two questions—"Oh, yeah?" and "Says who?"—and listen.

One day I challenged Patty's claim about the disease that had killed Max Hammond. The same sickness was killing other people connected to Cenex.

Oh, yeah?

Patty told me she'd talked to a teacher at the junior high school about his breathing problems, and he'd given her permission to call his doctor, who told her about seeing an unnamed environmental engineer from Quincy; that had to be Dr. Max. (I later confirmed with his family.) And Patty heard a woman who drove a potato truck was having lung problems, too.

Three people from Quincy saw a specialist in a rare lung disease— all connected to Cenex. The teacher's classroom was three blocks from the rinse pond site; the truck driver lived by a storm sewer drain that carried Cenex chemical spills; and Max was their main man. It was a cluster of disease.

Says who? As if Patty Martin would know a cluster.

A world-class expert, she shot back.

I'd check it out.

SEATTLE TIMES NEWSROOM

She was right.

Ganesh Raghu, a medical doctor and associate professor at the University of Washington Medical School, was one of the top researchers in the world on idiopathic pulmonary fibrosis. The disease scarred the deep tissue of the lungs until patients could no longer breathe. The prognosis was death, usually within ten years.

He told me, "There were all of a sudden four or five patients that got referred to me for idiopathic scarring of the lungs from the Quincy area. They were all in their middle-age situations. It is a cluster, it seems to me."

The statement surprised me, but Dr. Raghu was qualified to make it. He saw about twenty patients a week from across the nation. The disease usually strikes people in their sixties and seventies, not middle-aged people like the patients from Quincy, he explained. It afflicts fewer than one in five thousand Americans in a lifetime, not five from a small town in a few years. Yes, he said, it was definitely a cluster.

The cause was, by definition, unknown. The word "idiopathic" derives from the Greek *idio,* meaning "peculiar," and *pathy,* meaning "illness."

But Dr. Raghu told me, "Environmental factors are always implicated in any unknown disease as fatal as this one."

By now I thought the water group had something of importance not only to Quincy, but to the nation and world. I came to some fundamental conclusions. These were statements that were provably true. I wrote them in a notebook:

1. *Hazardous chemicals were being disposed of by turning them into fertilizers. While zinc was a minor plant food, the others, arsenic, cadmium, chromium, dioxins, lead, and mercury, came along for the ride.*

2. *Most people had no idea this was being plowed into the soil.*

3. *Of the few who did know about it, no one kept track of how much was happening, or where, or with what effect. Apparently it was increasing.*

4. *No one tested the fertilizers for the toxins. The United States had weaker controls than Canada and Europe. Industry ran the regulations. The only safety standards that existed were the ones for landfills and sewage, not for inorganic chemical fertilizer.*

5. *A larger and even more unknown group of industrial wastes was being put on farms under the rubric "bulk soil amendments."*

6. *It saved industry a lot of money in hazardous-waste disposal. It might or might not save farmers and gardeners a little money.*

7. *The soil is the start of the human food chain. Plants do absorb the toxics, especially cadmium, from the soil to the edible tissues.*

8. *There was some risk to human health. It was unknown and unpredictable how much peril there was.*

9. *Hard to believe? Yes. But the scientists, the regulators, the people inside the industry, all said so.*

Once I could frame the questions, the answers filled in the picture. It wasn't pretty.

I called the top regulators in the top farm states.

- In Florida, Dale Dubberly told me, "There's a lot of materials out there that have plant nutrient values, but nobody knows what else is in them."
- In Texas, George Latimer Jr. said he lacked authority to make the manufacturers reveal their ingredients not listed on the label.
- In Pennsylvania, Earl Haas said he tried to persuade industry not to put some waste in fertilizer based on the federal standards for sewage sludge, but in the end, he lacked legal authority over heavy metals in fertilizer.
- In Ohio, Bill Goodman said he noticed a lot more businesses distributing industrial by-products as fertilizers.
- In Kentucky, David Terry said, "Even traditional fertilizers might have some heavy metals in them. We haven't looked for a while." Terry said he wanted to start checking for hidden ingredients. Kentucky was one of only five states with fertilizer regulators based in universities rather than political capitals, inoculating them from industry pressure. The others were Indiana, Missouri, South Carolina, and Texas.
- In New York, the state fertilizer office, decimated by budget cuts, had even failed to check for guaranteed nutrients as a consumer protection measure, much less the hidden ingredients. The failure to check violated state law.
- Delaware, Illinois, North Carolina, and Pennsylvania regulators all said the use of industrial by-products in fertilizer was a major concern to them. Some others said they did not share the concern. They said it was beneficial recycling.

Canada was the only nation on earth that watched out for a palette of toxic chemicals in fertilizer.[7] Some European nations and Australia limited the amount of one heavy metal, cadmium, in fertilizer even tighter than Canada, but they never looked for the other toxic tag-alongs.

Canada limited nine toxics in fertilizer to amounts that would no more than double the background level in soil over forty-five years. Each product had a unique but specific limit depending on the recommended application rate. Canadian law also required anyone selling waste or micronutrients in fertilizers to provide four samples every six months, to identify all known contaminants, and to cite scientific reports focusing on food and environmental safety and worker protection.

"Yes, we support recycling, but only to the point that we do not compromise our agricultural crops," Darlene Blair of Agri-Food Canada told me in a phone interview. "In Canada, we've done a lot of work on industrial wastes and heavy metals to agriculture for a long time. We won't compromise our limits. Sorry, we won't compromise our health."

She continued in a voluble mood. Here was an area in which Canada was superior to all.

"In the U.S. we heard, 'Okay, how much can we apply until we get to the maximum that people can stand?' I was, like, 'Wow!' I think we're a little beyond the point where we wait till something is bad until we fix it. . . . The Department of the Environment is congratulating people for recycling things without understanding what the problems are with the recycled material. It's like, 'Way to go! Way to recycle!' And I say, like, 'Excuse me, but that recycled stuff isn't good for us. That's our food material.' "

In my own state of Washington, a soft-spoken, bearded, former South Dakota State University professor and poultry expert, named Ali Kashani, was in charge of fertilizer regulation. He said no state had a law setting a limit on the level of heavy metals that can be in fertilizer.

"That would be nice to have," he added almost wistfully.

Kashani relied on fertilizer producers to document their own safety. The state never checked back for toxic components. When Kashani wanted a scientist's view, he turned to Shiou Kuo, a research agronomist at Washington State University. Kuo was, if possible, even more soft-spoken than Kashani. Kuo supported recycling, of course. He told me most of the toxic metals are already in the soil and added in such small amounts as to be safe. But at some level, some unknown level, it would make people sick.

"This is something that troubles my mind," Kuo said. "Deep down in my heart, I think the less amount a toxic substance like cadmium is in the soil, the better. But in reality the question is really how much input can be tolerated. Until we know what the critical level is, this kind of question cannot be answered. So there will be a fear associated with the current practice. I just don't know if it can be avoided."

QUINCY, WASHINGTON

Patty drove down to the railroad tracks and noticed things no one else did. It was one of her favorite places to look and think. She spent more time than Glenn would want to know looking at the trucks and trains rolling into Quincy.

If she saw anything interesting, she hurried home and got the video camera and drove down again. She always took a wide shot followed by a close-up. Evidence. She had a particular interest in tankers parked behind the fertilizer plant.

On occasion, she took pictures of schoolchildren walking in the dust by the companies.

The Quincy police knew her activities well. The chief's office was next to the mayor's smaller room at the town hall. The chief had heard Patty's spiel. He'd been enlisted to help her decipher truck registration numbers and railroad product code.

At the mayor's request, the chief asked a trucker what he was hauling. Sulfur and other compounds, the driver said. Perfectly legal.

Other times, the tankers on trucks or trains would be missing the placards that identified their contents. Patty brought Bill Weiss down to look at an unmarked railcar that had gone off the track by Quincy

Farm Chemicals. She thought it was hazardous waste. It turned out to be a refrigerated car.

Another time, she chased after a truck with a placard showing it carried waste oil. It turned out to be carrying food-grade waste oil. Still, she was suspicious of every truck she saw.

Today she studied a train parked on the tracks. Two of the railcars were marked "Cominco Fertilizer." She caught her breath. She rushed home to call me. The railcar markings used to say "Cominco"—the name of a smelter in Canada—but now they said "Cominco Fertilizer."

Patty was quite sure that meant the company trafficked in hazardous-waste fertilizer to dispose of its lead and zinc leftovers.

She told me, "Somebody from the railroad said everything comes through Quincy—everything!"

I thought she exaggerated, but she was uncannily right about some things.

After that, Patty went out of her way, almost every time she drove a child to the school or the gymnasium or the swimming pool, to cruise by the chemical companies that lined the tracks.

She drove by slowly. She liked making them nervous.

SEATTLE TIMES NEWSROOM

Thank heaven for librarians. Our newspaper librarian, Vince Kueter, spent many days with me looking for reputable studies of toxics in fertilizer. We found oceans of science on toxic metals in sewage sludge, but only a spoonful of science on toxics in the common fertilizer. We found how much we don't know—the most important finding of all. This is what we learned.

One scientist did most of the work in America. He was John Mortvedt, "Mr. Micronutrient," the Borax scientist and Colorado professor whom I had met at the Heavy Metal Task Force in California.[8] Mortvedt had worked years for the Tennessee Valley Authority in Muscle Shoals, Alabama, at a national fertilizer research center allied with industry and excised from the federal budget in 1995. He was the coeditor of *Micronutrients in Agriculture*.

His first published paper on the bad actor cadmium in fertilizer focused on cadmium in the natural phosphate ore mined from the earth. He found the cadmium didn't just stay in the soil. At some levels, corn absorbed the cadmium. That occurred with phosphate fertilizers containing thirty to fifty parts per million of cadmium, a level reached by about 10 percent of the American phosphate fertilizer sales, particularly those from the phosphate mines of the West.

Another Mortvedt study found lettuce, peas, and radishes rapidly absorbing cadmium from fertilizer, especially on acidic, sandy soil.

In 1981, Mortvedt first wrote about industrial wastes carrying cadmium contaminants to farm soil. He said it was important to study because the cadmium could enter the human food chain. It was five years after Congress cracked down on hazardous-waste dumping. Mortvedt believed that recycling, done correctly, helped industry and agriculture alike. He was trying to find an acceptable level.

He found complexity beyond man's ability to predict. The variations were almost endless. Mortvedt found the toxic behavior of cadmium applied with industrial waste to the soil would vary by the type of soil, type of plant, and even by the part of the plant that was eaten. For instance, the bran, or outer coating, of wheat absorbed more cadmium than the endosperm inside the grains, in all the fertilizers Mortvedt tested. That meant you would expect to find more cadmium in whole wheat flour that keeps the bran than in white flour.

But Mortvedt, making some rough calculations about the American diet, concluded the cadmium levels in wheat probably wouldn't pose a health problem.

I did not think Mortvedt was qualified to judge health effects; he was a soil scientist, not a toxicologist.

I noticed the reference to bran, the first I'd seen of toxic chemicals targeting the whole wheat foods that have come to symbolize healthier eating. How ironic.

In another section of the study, Mortvedt said he would expect more cadmium infesting the soil around many fruit and vegetable crops because they take a lot more fertilizer than wheat. How much the fruits and vegetables absorbed would depend on individual soil chemistry and plant biology.

Swiss chard is a leafy vegetable, similar to lettuce, and a hyperaccumulator of heavy metals. Mortvedt tested Swiss chard and corn for a 1985 report, "Plant Uptake of Heavy Metals in Zinc Fertilizers Made from Industrial By-Products." It was the only published science we found on target. The results showed that chard would soak up cadmium put on the soil from toxic waste, while corn would absorb the cadmium at a lower rate.

Mortvedt said liming the soil would reduce metals absorbed through roots. This, he said, meant corn should be safe from recycled-waste fertilizers as long as the pH was kept up. And he said a lot of the cadmium would be left in the corn leaves and cobs not eaten by people.

In my view, Mortvedt's paper gave reason to let industry put cadmium-laced waste on corn fields. Corn accounts for about half of the twelve billion dollars in fertilizer sales in the United States each year.

As for leafy vegetables like Swiss chard, Mortvedt said they were not normally fertilized with the zinc fertilizer products most likely to contain hazardous waste, so he thought it should be okay. He summed up by saying that as long as the pH was kept up, the use of fertilizers from industrial by-products, in his opinion, "should not be a major concern in vegetable crop production."

To me it seemed Mortvedt was trying to justify the practice.

And what of the added toxics building up in soil over time? In 1987, Mortvedt tested plants and soil from nine long-term soil fertility experiments in Alabama, Illinois, Missouri, and Oklahoma. He found cadmium from the fertilizer slowly building up in the soil. In one case, Mortvedt estimated it would take nearly three hundred years for the level of cadmium in the root zone to double—statistically significant but in his opinion "negligible" in effect.

Other studies in Europe and Australia, in later years, found more of the fertilizer cadmium building up in soils and entering food crops.

European and Australian scientists expressed more concern about cadmium in the human food chain than Mortvedt did. As a result, the European Commission and Australia had placed strict limits on the amount of cadmium in fertilizer and food.

In his latest study of solubility of industrial wastes used in fertilizer, in 1992, Mortvedt demonstrated more caution. "Results of greenhouse experiments have shown that heavy metal uptake by most crops from these zinc fertilizers is very low," he wrote. "However, effects of long-term use of such products as zinc fertilizers are unknown."

To me, the unknowns dominated the landscape.

The other American scientist whose name always came up was Rufus Chaney.[9] He worked at the U.S. Department of Agriculture's Environmental Chemistry Laboratory. Chaney was an expert and industry consultant on phytoremediation, or the use of plants to suck poisons out of polluted soil. He had published more than 270 papers since 1969. His specialty was cadmium in the food chain.

Chaney agreed with Mortvedt about the low risk from the hazardous-waste fertilizer products Mortvedt tested in 1985, but for a different reason. It was the high zinc-to-cadmium ratio that Chaney said was the key. Zinc inhibits the absorption of toxic cadmium through the intestine. If the ratio of zinc to cadmium is at least one hundred to one, Chaney had found, the zinc would prevent the cadmium from causing harm. On the other hand, a zinc deficiency allows the cadmium in. It was Chaney's guiding principle.

In 1978, farmers in China and Japan reported severe illnesses from cadmium poisoning, known as itai-itai disease. Chaney found people near a former smelter in Pennsylvania eating potatoes from soil with fifty times higher cadmium than the Asians, but perfectly healthy. The difference, he said, was in their diets. The subsistence farmers in China and Japan were already suffering from malnutrition, deficiencies of calcium, iron, and zinc. As a result, their bodies absorbed more of the cadmium they ate.

Chaney believed some common fertilizer products might pose an eventual health risk to Americans and ought to be monitored better, disclosed to buyers, and cleaned up. He was thinking about cadmium levels in phosphate from the western United States, some as high as one thousand parts per million, which would be prohibited in Europe.

"We can't put that off forever," he told me when we discussed his work. "Europe has taken an option on all the low-cadmium ore that anybody can get in Florida and elsewhere. Idaho phosphate will become more important in the U.S. I could still argue there's virtually zero lifetime risk, but I would not be as sure.

"Manufacturers are trying to make sure they don't cause a problem or at least get caught causing a problem, so they're steering it away from sensitive soils like vegetables. They can put it off for a while, but not forever. I keep telling them that."

As for the micronutrient fertilizers, Chaney thought some were unsafe. A product in California had three thousand parts per million cadmium. He said he wouldn't put that on his garden, and farmers shouldn't put it on food crops.

In the 1970s, the Food and Drug Administration said Americans were consuming cadmium at a rate too close to the maximum recommended by the World Health Organization. The main source, for non-smokers, was grain and cereal products. Small, repeated exposures to cadmium can cause kidney and liver damage, bone disease, emphysema, anemia, renal dysfunction, and hypertension, as well as cancer and reproductive damage.

In 1993, Chaney drew on a national soil survey to publish a paper on toxic metals in the agricultural soils of the United States. He found cadmium highest in southern California coast ranges—anomalously high in some lettuce and spinach from the Salinas Valley—and in some vegetable soils of the Great Lakes states and parts of Florida, Oregon, and Idaho.

The heavy metals, he wrote, "persist in surface soils for centuries to millennia in the absence of erosive loss. The persistence in surface soils is the very reason why soil contaminants are of such concern to agriculture."

Still, Chaney told me when I phoned him, "In the U.S., we haven't made any food cadmium limits because we don't have any evidence they're needed by anybody in the U.S., and it would cost a hell of a lot to do the monitoring if only part of the crop is endangered. The FDA and EPA have concluded it is irresponsible to create unnecessary limits that cost a hell of a lot of money."

Chaney supported recycling of hazardous wastes to fertilizers as long as the contaminants were removed. "That's a responsible choice," he said.

But it wasn't always being done. The devil was in the details. And when the contaminants weren't removed from waste, even Rufus Chaney was concerned about using recycled fertilizer.

He was bothered by a case in southern Georgia. A mixture of steel-mill waste and limestone was given to peanut farmers as fertilizer. "Well, that wasn't a good idea," Chaney said. "When the farmers let the pH down by not liming later, it killed the peanuts."

So that was where I looked next. This was what I learned.

More than a thousand acres of peanuts in Tift County, Georgia, had been wiped out by heavy metals in fertilizer.[10] It was the worst confirmed case in America of fertilizer actually destroying a crop aimed for human consumption. The peanut plants were killed by a toxic brew of hazardous waste and limestone that had been sold—legally—to unsuspecting farmers.

Yet there had been no publicity. I could not find anyone who had checked the land, the crops, or the peanut butter for the hidden toxics.

The Georgia farmers didn't want to talk about it. But Jessica Davis, a soil scientist, did.

Professor Davis had been called in to help three of the peanut farmers detoxify their topsoil. The material sold as Lime Plus was laced with cadmium, chromium, lead, and zinc from industrial waste. She said it was a very sensitive situation.

"These guys were trying to improve their land," she said. "They weren't trying to hurt it. Now they're afraid if people know they had this problem on their land, they won't be able to sell what they've grown there and they won't be able to sell their land, either."

In Tift County, a three-hour cruise down Interstate 75 from Atlanta, farmers grew peanuts, pecans, and corn. They used manure, ground-up eggshells, and drywall to add organic matter to the soil. The ones who bought the black liming material from SoGreen Corporation were trying to save money on their farm chemicals.

Herman Parramore Jr. had opened a fertilizer factory on Maple Street in the county seat of Tifton. He got a permit from the Georgia Department of Environmental Quality to turn waste into fertilizer. Five steel companies paid the fertilizer man twenty-five dollars a ton to take the gray dust cleaned from their furnaces. The steel companies would have had to pay fifty, one hundred, even two hundred dollars a ton to store it in a hazardous-waste landfill in Alabama or send it to a high-temperature recycling company in Texas. They saved money getting rid of one hundred thousand tons for fertilizer.

Parramore used neither high temperatures nor high technology. He didn't have to. He just mixed a little natural limestone with the toxic ash and called it a fertilizer lime. The material was extremely alkaline—the selling point as lime substitute—and it was also a whopping 3 percent to 6 percent lead and dangerously high in cancer-causing cadmium, hexavalent chromium, and organic hydrocarbons.

The fertilizer company's field men told farmers their peanuts, pecan, and corn crops needed liming. A lot of liming. Two tons to the acre.

SoGreen spread steel-mill waste as fertilizer between 1982 and 1987, then went broke. Parramore blamed a slowdown in pecan growing, rising insurance rates, and red tape.

The peanut crops would not die until several years later when some of the farmers stopped liming. Then the heavy metals mobilized like an invading army and hit like a time bomb.

Davis was working at the Coastal Plain Experiment Station for the University of Georgia. She told the farmers the zinc killed their peanut fields. Peanuts are particularly sensitive to zinc. In a way, that was fortunate. The zinc poisoning kept even worse toxic chemicals off the dinner table.

A cautionary tale, Davis said.

"Let's say you're planting a crop that's not sensitive to zinc and so it doesn't die," Davis said. "Well, this material is high not only in zinc, but it's got lead and cadmium and chromium—all kinds of fun stuff that could be hazardous to humans. Somebody uses this on their sweet corn and eats it: Nobody knows how much lead they'll be eating."

The abandoned pile of fertilizer at the SoGreen factory was blowing a toxic dust over the low-income neighborhoods of Tift. That made news. The steel companies took responsibility for cleaning up the site after state officials threatened them with Superfund action. The companies proposed putting a five-foot-thick cap on the pile. The state said no. The companies proposed mixing it with cement and trucking it to a lined, hazardous-waste landfill. The state said yes. The cleanup cost more than ten million dollars.

Herman Parramore complained he could have moved the toxic mountain himself if the state had allowed him to sell it or give it away to pecan farmers. Parramore told a local newspaper, "I don't know of a case when this (dust) hurt anybody."

But the fertilizer maker had painful legal problems. He had violated his permit. He kept thousands of gallons of used nitric acid in a waste pit (making Patty Martin and Tom Witte think it sounded a lot like Quincy). In 1995, Parramore pleaded guilty to two felonies under environmental laws.

When I talked with Jessica Davis later, she had moved on to a teaching position at the University of Colorado. She said Lime Plus was still being sold in fifty-pound bags in Georgia.

Georgia officials denied that. In fact, Charles Frank, director of the plant food division of the Georgia Department of Agriculture, said he had no knowledge of any toxicity incidents at all in Tifton County, not then, not ever. Frank, a bald man known as "Curly" to his friends, said Georgia does a good job controlling industrial by-products on farms.

If you want to see real problems with industrial wastes going on farmlands, Frank told me, look at Alabama.

"Anything goes in Alabama."

QUINCY, WASHINGTON

She was relentless. She spoke on a cordless phone and paced like a big cat. Patty Martin quizzed a state official.

"Well, are you aware of any fertilizers or pesticides that have beryllium or cadmium or chromium in them?" she demanded. "Are you aware of any of those chemicals going into pesticides or fertilizers?"

Beryllium causes an incurable lung disease. Cadmium and chromium may cause cancer. Recent studies showed that even slight exposures could lead to sickness and death in people years later. They were found in the Cenex waste pond and had no place in fertilizer.

"Why didn't it send up a red flag when they were found there?" she pressed.

"Are farmers told that's in their product? Did we have any of that come this way? Where are they from? How did they get there? Why are they there? Do they have any value in agriculture? What health impacts do they have? Especially when it's windy. The beryllium was near our school. It's going out on farms."

The news never reached Quincy, Washington, when delegates from most nations of the world met in Basel and Geneva, Switzerland, to talk about the global trade in industrial hazardous wastes. This is what we learned later about the work of the United Nations Environment Programme:

The delegates in Basel in 1986 hadn't talked about fertilizer specifically, but they had raised many of the same questions as Patty Martin. Questions about using recycling as a sham or ruse to get rid of toxic waste on vulnerable populations. Greenpeace, which had grown stronger in Europe than in America, called the practice "the green mask of recycling."

On a global scale, the Basel Convention had agreed toxic wastes generally flowed from north to south, west to east, rich to poor, each party finding a short-term interest in paying or being paid to ship or receive the wastes. A majority of the 116 nations represented in Basel had signed new restrictions on international trade in hazardous waste. But the Basel Ban was riddled with loopholes. By 1994, the sham recycling issue was coming up for another big meeting of nations in Geneva.

By then, Patty and Dennis had met each other in isolated, faraway Quincy, started testing Cenex fertilizer, and contacted me, but none of us knew yet of the United Nations work.

The global waste-to-fertilizer trade had also gained limited notoriety in 1991 when three companies in the southeastern United States were caught disposing hazardous waste in fertilizer shipments to Bangladesh and Australia.[11] The companies were Stoller Chemical of Charleston, S.C., Gaston Cooper Recycling of Gaston, S.C., and Southwire of Carrollton, Georgia. Although the illegal trade under guise of fertilizing material resulted in criminal cases and a furor in Bangladesh, few people outside Greenpeace knew of it in the industrialized world.

We would learn, then, that by the time of the Geneva conference, Third World countries knew—or suspected—much more. They insisted on a total prohibition on toxic-waste exports from richer nations to poorer nations, no matter what business interests in either place said they were trying to accomplish, and yes, even when recycling was the claimed purpose.

Greenpeace had identified "the Sinister Seven" as Australia, Canada, Germany, Japan, the Netherlands, the United Kingdom, and the United States.

On the other side: the so-called Group of Seventy-seven, led by Devanesan Nesiah of Sri Lanka, who said the trade simply sent toxic waste "from countries that can cope, to countries that cannot."

While African, Caribbean, and Pacific nations had made some progress on their own in protecting themselves from their greedier short-term interests, they hadn't stemmed the tide. The World Conservation Union and Global Legislators for a Balanced Environment (GLOBE) had supported a total ban on waste trade, and Vice President Al Gore had announced White House support for a ban, at least beyond North America—which conveniently allowed the United States to continue its waste-recycling trade with Mexico. The United States Congress had never ratified the Basel Ban, and the U.S. Chamber of Commerce was stoutly opposed. The U.S. delegate at Geneva said the advanced nations needed the freedom to promote recycling and technology transfer to poorer nations.

This was how the poorer nations acted: China, then Hungary, Slovakia, Croatia, Slovenia, Ukraine, Poland, the Czech Republic, Estonia, Latvia, and Romania, added their voices to the Group of Seventy-seven. The underdogs won a two-thirds majority at the conference. When Nesiah announced they'd negotiate only on the starting date of a full ban on waste exports, a silence fell on the conference room in Geneva. According to Greenpeace, the delegates from the United States and other richer nations had not expected such opposition to what they viewed as legitimate recycling.

And then Western European countries joined the call: Denmark, then the Netherlands, Portugal, Belgium, Spain, France, Italy, and finally Germany and the United Kingdom. That left Australia, Canada, Japan, and the United States opposing the ban. The final language was hammered out late at night by representatives of those four countries and the European Union on one side of the table, and Sri Lanka, Antigua and Barbuda, the Bahamas, El Salvador, Colombia, Senegal, Egypt, and Poland on the other.

The decision prohibited immediately all movements of hazardous wastes for final disposal from the twenty-four richer countries of the Organization for Economic Cooperation and Development (OECD) to the less industrialized nations of the world. It prohibited the movements for recycling enterprises effective in 1998.

Greenpeace crowed, "After 1998, hazardous waste traders will no longer be able to justify hazardous waste exports by sending them to sham or dirty recycling operations in Eastern Europe or the South. For the first time in international law the Basel parties took a clear decision that hazardous waste is not a 'good' suitable for free trade, but something to be avoided, prevented and cured, like a disease or a dangerous plague."

But the ban would not restrict waste trade between OECD countries, such as the United States, Mexico, and Canada, or between non-OECD countries, or from non-OECD countries to OECD countries. And it would not affect paper, glass, scrap metal, plastics, household waste, or the ash residue from incinerated household waste, unless they were found to be contaminated with heavy metals or dioxins. Since that finding hadn't been made, those wastes were still going from rich to poor to recycle into fertilizer and other products.

Not long after the delegates in Geneva adopted the new rule, the wealthier nations started subverting it. The most pervasive new shams, according to Jim Puckett of Greenpeace and the Basel Action Network, were to redefine waste that was bound for recycling as nonwaste, and to redefine materials that were previously considered hazardous as nonhazardous. "In other words," he wrote, "rather than trying to eliminate the hazardous wastes, governments are trying to eliminate hazardous waste definitions."[12]

The worldwide political fight held valuable lessons for Patty Martin when she found out about it later.

"This is so huge!" she exclaimed many times.

But it was no help at the time and place Patty needed help the most—home in Quincy.

MARCH 1997 — QUINCY, WASHINGTON

Tony Gonzales got a copy of "Lead in Your French Fries?" and took it to work. Dwight Gottschalk was apoplectic.

The paper took direct aim at the company he managed. Lamb Weston made french fries twenty-four hours a day. The Quincy plant sold mountains of fries to McDonald's. Was Tom Witte trying to set off a food scare? Ruin their business? Drag them down with his miserable farm? Four hundred and fifty people worked at Lamb Weston, including Glenn Martin.

And look—a letter stapled to the back of "Lead in Your French Fries?" talking about Mayor Patty Martin complaining about toxics in Quincy at a conference in Washington, D.C. Going around with Nancy Witte causing trouble. Confronting the head of the EPA.

Dwight and Tony both laid into Glenn to try to control his wife.

"*You* try to control her," he said. "I can't."

Glenn did not tell the men that his wife had already made plans to stir up more trouble. She was holding an airplane ticket to go to the second national conference on children's health, with Nancy Witte, in Salt Lake City.

Glenn knew Patty didn't agree with some of the speculation in "Lead in Your French Fries?" The Wittes had asked her to read it in advance and she'd declined. Glenn had told her to keep her distance. The Wittes, bankrupt, were practically immune from being sued if they slandered someone. The Martins were not immune.

Glenn knew Patty had asked the Wittes to leave her name out of the paper. Then they'd attached the letter talking about Mayor Martin going to Washington, D.C., with Nancy Witte.

But Glenn was mad at Tom and Nancy, not Patty. He understood Patty. She had to do this. When Patty needed him, he came through. Glenn didn't lead the charge, but he didn't run the other way. When Patty felt under the harshest attack, Glenn told her she was doing the right thing. They'd been together so long. Glenn knew Patty had a good heart.

But Patty Martin was not a popular politician in Quincy these days. She was shunned by old friends and threatened by other business leaders.

As mayor, it was her business to worry about the public health. The Grant County Health Department agreed to do a health survey. But there was a problem. The Quincy weekly newspaper didn't mention a word about the survey. Patty, Nancy, and a school nurse were the only three people to show up when the department held a public meeting.

So Patty wrote the state health department. They agreed to come to Quincy.

Dennis, meanwhile, was stopping at courthouses and libraries during truck runs all around the state. The onetime "Farmer's C" student was looking up court cases and statutes. Dennis found a section of the law allowing the director of the Office of Waste Reduction to accept gifts and grants from private businesses. He and Patty thought, *Legalized bribery.*

Patty sent a letter to the Office of Waste Reduction asking for a list of all the gifts it had received from companies. The office manager replied they hadn't taken any.

Word got back to Quincy city council members that Martin was falsely accusing state officials of taking bribes.

They were unhappy with their mayor. They heard she'd been consorting with the Wittes and talking with a Seattle newspaper.

She talked too much, they thought.

She was out of control.

It was time to rein her in.

Jerry Eide, an engineer who took the late Max Hammond's job with Cenex, told Patty he wanted to dig up the top six inches of soil at the rinse pond site and haul it away to a licensed landfill. He thought that would solve the problem she was complaining about. Patty said no. She thought it would destroy evidence.

Eide told her she was playing with Goliath. Bad analogy, she replied. David slew Goliath.

Patty made flurries of phone calls from her desk in the laundry room. Her baby had heavy metals in his hair. Her friends were sick.

Tom's cows and Ruthann's horses died. Nobody would tell them what was in the fertilizer.

She thought the Cenex pond had the proof. Patty tracked down somebody new, a fill-in manager at the state department of ecology. He listened to the mayor's plea and phoned Eide and told him to leave the dirt there.

Within minutes, the phone rang at the Martin house. Patty answered. It was Pete Romano. Patty didn't bother sitting down. She knew it would be a short call because Pete and she frustrated each other.

Pete's company, Quincy Farm Chemicals, was one of the few fertilizer makers left that wasn't owned by a conglomerate. He was a small fish. He thought he was vulnerable. And he would not tell Patty this, but he did use some Bay Zinc hazardous waste in fertilizer. Pete believed recycling helped the environment.

"What's better, put it in one big hole?" he asked later. "Or to process it and put it into a form that is usable and spread over many, many acres at very minute levels? What makes more sense?"

Now Pete had a simple message for the mayor. He said they had to get the Cenex dirt out of town and put it in a landfill. *Move on.* It wouldn't be wise, he told her, for a mayor to side with a few broke farmers and hurt everybody else. "What we have in Quincy," he said, "is three broke farmers that got into this heavy-metal kick as an opportunity to look for some deep pockets."

Most farmers were doing very well, thank you, with the fertilizer and the help and advice they'd received from the chemical companies. Patty ought to help them, or they'd hurt her.

She got the message. She replied tersely:

"Good-bye, Pete."

He put the phone down hard.

Romano and Eide, Gottschalk and Gonzales, John Williams and Nick White, the fertilizer lobbyist Scott McKinnie in Spokane and the food products lobbyist Craig Smith in Portland, were burning up the phone lines with each other.

A big day was coming. A bushel of state officials would meet with the mayor and council and citizens in Quincy. A reporter was nosing around. This could explode.

Patty Martin was blowing this up, and who knew where it would stop? They came up with a plan to stop her.

CHAPTER 7

POWER AND PROOF

April 22, 1997

Ladies and Gentlemen:

We are concerned with the activities of Mayor Patty Martin. She has identified herself as Mayor Patty Martin of Quincy *and has made allegations that dangerous wastes and toxic materials are used as additives in soil amendments or fertilizers. Patty Martin has traveled to Washington D.C. and attended a children's health conference voicing her concerns regarding additives used in the Quincy area. She claims to have data showing that toxic chemicals and toxic heavy metals were used in fertilizers and soil amendments applied in the Quincy area.*

The EPA has requested that Mayor Martin send in the data so that it can be studied by the EPA. To our knowledge, she has not done so.

It appears that the concerns and allegations made by Patty Martin have no support. There are numerous ongoing studies on the ingredients in fertilizer. Cenex, for example, has volunteered to have its fertilizer tested by an independent agency. Both the [U.S. Department of Agriculture] and [Washington Department of Ecology] have ongoing studies.

The problem with Patty Martin's activities is that her name is now synonymous with the Mayor of Quincy. Any actions that she takes are actions of the City of Quincy.

The apple industry has barely recovered from the Alar scare which cost the industry billions of dollars. If the unfounded, baseless allegations *of Mrs. Martin continue, the Quincy area could suffer drastic economic consequences. A group of farm operators and food processors are prepared to file a lawsuit against the Mayor and the City if these unfounded allegations continue. If people and/or food distributors decide not to purchase food products from this area, the City of Quincy and the surrounding area would be economically devastated.*

In the 1930's Quincy literally dried up and blew away. We respectfully submit that the unfounded allegations regarding additives to the soil could result in an economic devastation similar to the 1930's.

If there is something wrong, let's make sure that it is fixed, but let's not destroy this area for somebody's personal political gain. This letter is to put the Mayor, the City Council and the City of Quincy on notice that we will seek to recover damages incurred as a result of the notoriety from the unfounded allegations that are being made.

Very truly yours,
Bill Weber
3559 Road K N.W.
Quincy, Washington 98848

The letter was only the beginning of the heavy-handed attempt to silence the mayor. The community powers in farming and food and chemicals and politics and law were also lined up against her. Threatening a lawsuit accomplished two things. It warned Patty Martin she'd be held responsible for financial losses. And it allowed the town council to take her behind closed doors, where everyone agreed with what one of the men yelled at her:

"I told you! *Be quiet!*"

And where they told Patty exactly what she had to do.

———

Quincy Town Hall was closed for two hours while I waited outside. Then the doors swung open and the council members and mayor walked quietly out to their cars under a sky near twilight and translucent blue.

Patty, exhausted, drove home and drank a glass of wine. She told Glenn she'd tried to keep her mouth closed in the closed-door meeting. "You get more information by sitting back and listening," she said.

She'd already known the plan. A council member had tipped her off. And it was just as harsh as she'd expected. The council and city attorney had told her to choose: Stop talking about fertilizer, or step down as mayor. If she wouldn't choose, the council threatened to strip away her legal indemnification as a public official so the Martin family would be personally liable for any damages in court. They said that if Patty caused a food scare while she carried the title of mayor, the town itself could be sued, and they would not allow that.

Glenn told her, "Don't let them buffalo you."

———

Before turning off the lights and closing the town hall for the evening, Sue Miller, the city clerk, typed up a notice, faxed copies to two newspapers and one radio station, and posted a copy on the front door. There would be another council meeting the next day. The citizens of Quincy would find out what their mayor would do.

LEGAL NOTICE

The following is the agenda for the Special Meeting:
1. Concern of Citizens as they relate to action or inaction taken by

*the Mayor or City Council regarding hazardous and/or toxic condi-
tions existing in and around the City.*

*The Special Meeting is open to the public and they are encouraged
to attend.*

———

The council chambers wouldn't hold all the people who showed up,
not even close. More than 150 people, double yesterday's crowd, overflowed the town hall as twilight washed over Quincy. They needed a
bigger place. So they all walked or drove six blocks down elm-lined
D Street, past the Faith Community Church and Royer's Home Fur-
nishings, past U.S. POST OFFICE QUINCY 98848 and the *Quincy
Valley Post-Register*, past Barb's Clothes and Rob's Video, to arrive at
the canopied sidewalk of Paddy McGrew's Restaurant and Lounge. It
had a big room in the basement.

Go ahead, the owner said. Use the basement. But he was going to
lock the front door to the restaurant at nine o'clock, so they'd have to
leave through the lounge. That was fine. Some people might want a
drink anyway. They settled into chairs and stood along the basement
walls.

The meeting started at half past eight. Comments, three minutes
each. Council members first.

George Nutter was aggravated as he talked about "Lead in Your French
Fries?"—"a simple phrase tweaking everybody in this community"—
and the letters Tom and Duke and Patty had written to the EPA and the
state departments of ecology and health.

"Nothing happened!" he said. "It's all letters with no action. And
here we are tonight adding credence to these unfounded accusations.

"I've been here eighteen years and in my opinion this is, outside of
Mount St. Helens, probably one of the biggest catastrophes that I've
been through. But what separates us from Seattle, Bellevue, and the rest
of the communities is the fact that this community pulls together. If we
start fighting amongst ourselves and if we keep directing attention to
this, they win."

Nutter looked at the crowd.

"Take the opportunity to vent. Don't hold back. Express your-selves," he said. "But remember it's Quincy, and we are a community. We want to get through this, but we won't get through this fighting amongst ourselves. We'll get through it when we pull together. When the city and the county residents and the farmer and industry, when we pull together, we are the voice. We make the rules. As a community. That's all I got to say."

There was loud applause.

Greg Richardson, a young silver-haired businessman and the presi-dent of the state potato commission, stood, poised, coiffed, and as debonair as a prince in court.

"I think the people of this community need to know at this point what position the mayor is going to take in regard to this, whether she's going to step back from this issue, whether she's going to pursue it, because this issue is being targeted at this community through her activities and this is where the detriment is coming," he said.

"She has misused her office to get into higher places to voice her personal concerns and I think this is totally inappropriate.

"There's a couple options, I think: One of them is quit talking about it. The other is step down from office.

"This room is full of successful farmers, people that haven't gone bankrupt because of fertilizer. We all have used these fertilizers, they are good fertilizers, they're supplied by good people, and we have been successful using them. We will continue to be successful. To go bank-rupt, to lose your farm takes a lot of time, lot of years. It don't happen overnight. And it don't happen because of fertilizer."

The crowd broke out in applause again.

"Now, Patty, you know we got an ugly situation here," Richardson went on. "This whole community is at risk. And the proper authorities are dealing with the situation already. There is no need to bring it into the news. There's no need to pursue it from Quincy. They're doing the studies in California right now. Why do we need to take our issue any farther then it's already gone? I think from what you've heard from last night's meeting and tonight's meeting, that we *do not want you* to pursue this in our interest."

The crowd roared. Patty felt numb. She searched out Glenn's eyes for kindness in the crowd.

One after another, members of the crowd in the basement of Paddy McGrew's stood and agreed with Nutter and Richardson.

The mayor needed to think about Quincy.

She needed to help the chemical companies who helped farmers.

She needed to avoid a food scare at all costs.

From the back of the room after a while, Duke Giraud stood and walked forward to the microphone.

"I'm probably one of the 'popular' people here tonight that Greg was referring to." He looked over the crowd. "I'd like to know how many of you guys have researched this yourself," Duke asked. "How many have checked into waste going into fertilizer? Has anybody?"

Somebody yelled, "That's what we have the department of ecology for!"

"Well, no, I didn't ask that. Have you checked for it?" Duke asked.

"I don't need to!"

"Okay, you haven't checked. Has anybody else? You know, when the department of ecology checks fertilizer, do you know what they look for? Nitrogen and phosphorus and potash; N, P, and K. That's it. They don't look for anything else."

"Are you sure?" somebody asked.

"I'm sure. Ask 'em. The guys were here yesterday. Did anybody ask those guys? The toxic-waste reduction people? Do you know where that word comes from—toxic-waste reduction? Not just from Superfunds. Now hear me out, I got my three minutes, too. They can legally put waste into fertilizer. Good, bad, or indifferent. I'm not buying waste. I don't know who puts it in. I haven't learned that yet. But I'm just saying, this is a real issue."

"What have you found out?"

"That they're doing it."

Several people spoke up at once, and they were not congratulating Duke.

"Wait, wait, wait!" a council member said. "He has his three minutes."

"I'll be glad to talk after this meeting," Giraud said. "Let's get this meeting done and I'll talk to anybody that wants to talk about it. Anybody. This has been no secret. I'll get you documents about it."

"Who's doing it?"

"I—I'm not going to say who's doing it right now."

"HA!"

"I'm saying it happens. And you guys are saying, 'No, it doesn't happen.' I can prove to you it does."

"Well, then how many of our farms?" somebody said.

"Well, you know, that's neither here nor there. That's not what I'm saying right now."

More hooting and jeering. Duke thought they looked like a lynch mob in a Mel Brooks movie. He thought they were going to throw tomatoes at him.

"You know you guys all want to make fun and laugh and poke and make fun of me, huh?"

"You have something to tell us?" a man interrupted.

"I do. They're putting waste into fertilizer."

"Have you got documentation?"

"I've got documentation."

"Show us right now. We're all here ready to listen to you."

"I'm not going to. Not right now, Jim. I'll tell you after the meeting—but what's the matter, don't you want to talk after the meeting?"

Talking among the residents.

"Yeah, I didn't bring my documents, but I'll be glad to show you guys. Let's have a meeting."

More rumbling in the crowd.

"No, I'm not prepared—I didn't figure anybody'd even be—I've made my point."

Duke thought a moment and then added, "I want to talk to an attorney before I can start mouthing off about who's putting it in their fertilizer. Now once I find out what I can and can't say, I'll gladly tell 'em. They'll find out."

"Thank you," Patty Martin said. She felt bad for Duke, especially when he could have brought some paperwork with him but didn't. Patty thought it would do no good to defend Duke because it would further estrange her.

"Okay, are there others that would like to address the council?"

Duke ambled back and sat down next to Tom Witte.

Tom leaned over and said, "I'll go up there next."

"Save your breath," Duke muttered.

Lance Hammond was speaking at the front of the room. "I've read the letter," he said, referring to "Lead in Your French Fries?" "There's a lot in the letter that I'm not qualified to judge. But I can tell you there's also a lot in that letter that is pure garbage. And when you mix the garbage with whatever facts you might have, it's totally useless. Because nobody can tell then what's right and what's wrong from what you've written.

"The accusations that you've made about my brother, about other people in the community, are baseless. And they are wrong. And they hurt people, and as I said the other day, I'm embarrassed to live in a community where something like— someone would say such unfounded accusations. So I don't really care right now whether what you have is true or not. The way you put it together, all it's doing is causing a lot of contention and problems in the community. That's not helping anybody.

"If you have things that are true, then document 'em and publish 'em as documented truth. But don't make allegations that are not true and just cause a lot of strife for everybody living here. Thank you."

The applause filled the basement.

"Can I have a question?" Richardson spoke up. "Patty, I think the people are waiting to see what your action is going to be, uh, before they speak. And depending on your decision, we may walk away peacefully."

Laughter.

"Uh, you know, otherwise, I think there will be some more discussion."

"Okay, I guess." Patty sounded puzzled, unsure. "What I should ask is whether there is anyone else who would like to address the council, and if there's not, then—"

Bill Watson stood up. A former president of the potato growers group, he held a copy of "Lead in Your French Fries?" He had been a referee in Patty's high-school basketball games. He spoke directly to her.

"There's a lot of things that I get to thinking about and wondering, and one of them is, the action that you've taken under the guise of the mayor of Quincy. How much liability insurance do you have or does the city have?"

Ten million dollars, Martin said.

"I'm like Greg," Watson continued. "I'd like to stick around and see what the decision is and how you're going to pursue this, and then go from there. Because I think everybody's concerned."

Patty said, "I agree, Bill." She felt wounded by Watson's attack. She'd looked up to him.

Watson added, "I mean, hell, this is where nine-tenths of 'em in this room have grown up. This is where they make their living. From agriculture. Let's not give it a black eye."

"Thank you," Patty said.

The residents applauded.

Martin: "Any other comments?"

Another man spoke, "It appears to me, councilmen—I'll just address all the council—that our mayor is kind of a loose cannon."

Laughter. Applause.

They're always trying to put her down because she's a woman, I thought.

The council was about to talk about a resolution. Patty knew it would be a motion to shut her up. That's what it was all about. She asked if she could talk for a minute first. The crowd fell silent.

"I grew up in this community," Patty said, "and I have absolutely no intention of harming anybody, okay? Please don't judge me guilty until you have the facts. I went to Washington, D.C., with the best of intentions being concerned for the health of the children at the junior high school. Okay? I asked a question back there and I didn't even introduce myself as the mayor of Quincy. All right, I did introduce myself as a mayor of a small community in Washington State who has some health concerns—"

"When you go home and have some french fries tonight!" a man yelled out to laughter.

Patty looked up and saw a friend of Tony Gonzales's, grinning.

"I inquired because the first thing I thought was 'Does the farmer know?' 'Who's telling the farmer?' " she went on. "And I'm sorry if I presumed incorrectly that you might have wanted to know if there was something like this going into your fertilizer."

Her voice was tired. There were shadows under Patty's eyes. She hadn't slept well the night before. Patty had wanted desperately to nap prior to the meeting, but she didn't have time. Now it was past ten. She wanted to do one more thing and then go home.

"I would just like to read a letter I wrote today, okay?" The mayor pulled a piece of paper out of her purse. It was her decision.

Citizens of Quincy
Members of the City Council
For the record,
Regarding recent concerns over my participation as Mayor in the pursuit of knowledge regarding hazardous substances in fertilizer, and

specifically whether it is appropriate for me to be involved as a City representative.

I answer in the affirmative. My involvement is consistent with the City's Mission Statement to ensure the Health, Safety and Welfare of the Citizens of Quincy.

It is consistent with the Freedom of Information Act's policy of "people's right to know," and with the Federal Community Right to Know Act (an act which specifically deals with the community's right to know about the presence of toxic chemicals).

During the course of my participation to protect Quincy's health, it never occurred to me that I would encounter opposition to what I perceived as protecting the community's health, both physical and economical.

Recognizing that this pursuit is no longer consistent with a direction the City would now describe as "in its best interests," the city has chosen, with my support, not to pursue the issue of hazardous waste in fertilizer.

"Satisfied?" she asked.

Patty had written the letter an hour before the meeting. Glenn had encouraged her to write it.

Tony Gonzales made a motion: "Due to the fact the town of Quincy did not have the resources to research the issues of heavy metals in fertilizer, any information received by the City will be turned over to the proper agencies for their research and determination of safe levels."

Passed unanimously.

Meeting adjourned.

Nobody asked Duke to see his proof afterward, or the next day, or the next. He told Jaycie, "I really thought I could talk to a few people there who would be interested in this. Well, it looks like they're in charge of this. At least it's not a secret anymore. It's out there."

Larry Schaapman, the farmer and church leader, walked up to Glenn Martin while people were leaving the basement. *Your wife is a very strong woman,* Larry told Glenn.

The editor of the Quincy newspaper spoke to Glenn, too. *I'm sorry, Glenn,* she said, with a tear in her eye. *For what?* he asked. She didn't say. Later Patty would take it to mean sorry the newspaper had to print two pages of articles quoting the people attacking her.

Glenn went home to check on the kids while Patty stayed to talk with a few more people in the basement of Paddy McGrew's.

She felt tearful. She told Jerry Eide, the Cenex scientist, "You've got me."

Patty stayed hoping to talk with Dwight Gottschalk, in whom she had confided so much about the wastelands south of town. She asked Dwight to drive her home.

They sat in his car talking. Dwight was agitated. He'd done a lot for Glenn and Patty. Dwight thought her behavior was irresponsible. She wasn't seeing the big picture. Dwight wanted to know just one thing:

Who brought the reporter in?

At home, Glenn told Patty he'd never been so proud of her. She'd taken all that abuse and survived standing.

"I didn't know you could do that," he said.

Patty thought Glenn had taken a lot of abuse, too. What abuse he didn't get out in public, he got at his office in the food plant.

"I'm sorry for all this," she told him.

Patty promised Glenn she wouldn't go to the children's health conference with Nancy Witte—eating a $240 plane ticket—and she wouldn't talk about fertilizer anymore.

SEATTLE TIMES NEWSROOM

As soon as I sat down, the phone rang. Patty Martin again. Every day she'd call two or three times, usually with new and ambitious ideas of what we should be investigating, or excited news of this or that sufferer or scientist or conspirator. Her twangy voice would be confident and demanding. But this morning, she sounded slow, low, and groggy.

"I can't talk about hazardous wastes anymore," she said. "Anything I say about this, I could be sued for."

Silence on the line.

"I'm sorry. I guess I can't help you anymore. I couldn't even sleep last night thinking about all this. I was actually crying—Me!—I haven't cried since my father died."

I tried to shore her up. Free speech and all that. The First Amendment was still the law, as far as I knew, even in Quincy.

But Martin said she'd given her word. She'd written a letter promising that as mayor she would never again speak about toxics in fertilizer, and the letter was being printed on the front page of the Quincy newspaper that day.

"I was forced to write something," Patty said. "I just wrote it, then printed it out and took it to the council. I couldn't even look at it again."

I asked how she felt. Abused, she said, but relieved; pummeled but proud that she'd already played a big part in getting to the truth. Patty said she felt safer surrendering to the demand for silence because she knew I had joined the hunt, carrying the power of the press.

"This issue was too big for me to handle," she said. "It was outside anything the city could work on. My concern was the proximity of those fertilizer plants to the school. But it's a lot bigger than that. We know that now. Why else would they be so afraid of me talking about it?"

I was disappointed at her capitulation. But I said the newspaper would get the truth out. I didn't tell Patty, not then, how much I was relying on her as the whistle-blowing mayor at the heart of the story.

SALT LAKE CITY, UTAH

Nancy Witte was partway to the Second National Research Conference on Children's Environmental Health, driving her old pickup truck to save money, when Patty canceled her plan to fly over and join her.

Nancy couldn't afford a hotel room without Patty sharing the cost, so she pitched a tent in a campground outside the city. She cooked dinner over a propane stove, swearing to herself about Patty, paranoid Patty. The next morning, Nancy washed in a camp bathroom with a cold stone floor and changed from her usual jeans and T-shirt to a dress and drove into Salt Lake City alone.

There were many stars under the Utah sky: Olivia Newton-John, breast cancer survivor; Nancy Chuda, Children's Health Environmental Coalition founder; galaxies of activists and experts gathered at the Sundance Institute—few of them interested in Nancy Witte's far-fetched story.

But Nancy was used to being underestimated and overlooked. She was stubborn as a farm girl herding cows for milking, patient as a special-education teacher corralling wild children for class, both of which she did regularly.

The conference room was buzzing. Nancy set up a display board. She pasted a skull-and-crossbones drawing and a copy of the lab reports on metals in the Cenex rinse pond, Tom's fertilizer tank, and Duke's potatoes. She stood and talked to people and handed out "Lead in Your French Fries?"

Nancy thought she made progress, but needed more help.

She saw Lynn Goldman. Goldman told her the EPA had checked hazardous waste in fertilizer and found there were no problems. End of story. Nancy also met a man named Ken Cook, who was perhaps the one person there who understood her best.

Cook had tested a lot of dirt in his time. He was a soil scientist and president of the Environmental Working Group in Washington, D.C. The group supported recycling—*Of course!*—but Nancy showed him things he hadn't known.

He promised to help.

"I think it's an important story because we don't know where all this stuff is going or how much is going on the soils," Cook said later. "Fertilizer made from industrial waste with heavy metals may be an occupational exposure for farmers. It's clearly at least loading into the soils. And it may be getting into some of the crops.

"There's a logic to it, of course. This material lowers the farm chemical costs and avoids landfill costs, too. Those are patently true. Right now it appears there's an economic use of this waste material, but it may just mean that we haven't looked that far down the food chain yet. Sometimes it's a bonanza if it can be recycled and sometimes it's just a shell game where we're transferring the risk back to the land."

Could it be perfectly safe? Cook said, "Let's put it this way: We're well into the use of these materials before these questions are even asked. That doesn't seem to me to be a good sign that we've been very rigorous in our science on this.

"You can make the case under lots of circumstances that these materials won't be taken up in the food chain. But you can also see the pH might not be maintained. And even if it gets flushed out of the food chain, maybe it just takes longer to build up to the threshold effect. And maybe there is no threshold level, maybe it's linear, with things like lower IQ coming from any exposure to lead.

"The story of pesticides was to pick up on the warnings late. That was the lesson of *Silent Spring* and of Alar."

SEATTLE TIMES NEWSROOM

Calling it an Alar "scare" reflected a common misconception.[1] It wasn't a false alarm. Within a year of the CBS *60 Minutes* broadcast, Uniroyal, the manufacturer of Alar, voluntarily pulled its food use registrations, and the EPA banned its use in food. Alar was indeed a most potent animal carcinogen in our food supply and probably caused cancer in children. The apple growers who sued CBS lost their case at every level of the courts.

But they would not forget. Alar cost the growers an estimated one hundred million dollars, and some lost their homes. The eleven Washington farmers who sued CBS were from a town an hour's drive from Quincy.

Somebody leaked me this memo:

April 29, 1997
TO: All Processor Members
FROM: Craig Smith
    Vice President, Environmental Affairs
    Northwest Food Processors Association
SUBJECT: Potential Food Safety Crisis—Heavy Metals in Fertilizer
A small group of former growers in Quincy, Washington, have published a document titled "Lead in Your French Fries?" They claim that fertilizer companies have conspired with the EPA and the Washington Departments of Agriculture and Ecology to "hide" toxic levels of heavy metals in fertilizer. The mayor of Quincy has been heavily involved with these growers and has seemed to give them some degree of undeserved credibility.

The claims in the document appear to be unfounded and untrue. However, a reporter for the Seattle Times has picked up the story and is planning to write a series of articles on this topic. We believe he may attempt to tie the allegations to the possibility that the metals are being taken up in food crops, potatoes in particular, and that there may be a risk to children.

While we do not believe that there is any data to indicate a problem with heavy metals in food, we are taking this issue very seriously. The potential for a large scale, unfounded food scare is real.

The Northwest Food Processors Association covered seventy-one companies with combined revenues of six billion dollars a year. Almost half came from J.R. Simplot, the Idaho billionaire who started making fertilizer to feed his potatoes. They went on alert over the prospect of my newspaper story.

Craig Smith was a big, hearty man, young, well groomed, well spoken, quickly taken to be friendly. He was a hell of a lobbyist. Smith had a plan.

The food processors would work with state officials, fertilizer lobbyists, and growers. They'd test Washington potatoes. They reached out to a professor described by Smith as "an independent toxicologist" who would review the tests and say they posed no risk to health. That was the plan.

But Professor Allan Felsot was not independent, not a toxicologist, and not an expert on health risks in food. He was an entomologist, or bug expert. Felsot had been hired by Washington State University at the request of apple growers who'd been hurt by the reaction to Alar. Felsot's name was first raised in a phone call from Craig Smith to me. The friendly Smith wanted me to talk with the professor. I knew I was being lobbied. In all our searching for scientific papers on heavy metals in fertilizer and food, in my interviews with top scientists in the field, Felsot's name had never come up.

It turned out that Felsot was not volunteering his time, either, as Smith had told me. He was paid five thousand dollars by the food processors. Professor Felsot—a former Rhodes scholar with a ponytail—had a history of being paid large amounts of money to produce reports for his benefactors in the chemical industry.[2] I waited to see his report on toxic metals in food. I saw in a first reading that it was obviously slanted in many respects, and every slant went in the same direction.

So I set the report aside and sent a fax to Professor Felsot asking him, as a public employee in a state university, to tell me where his money came from.

Then I took a trip. Patty could be silenced in Quincy, but I'd heard of someone in Idaho who knew the whole inside story and loved to talk.

POCATELLO, IDAHO

Old John Hatfield lay on top of a bed, fully clothed, his window overlooking the brown town of Pocatello but his eyes focused on the hospital ceiling. He'd had prostate surgery earlier in the week. He was brooding, waiting for me, impatient to get out of there. And the staff at Memorial Hospital appeared to be at least equally impatient to get rid of the cantankerous old fertilizer salesman. No one said hello or good-bye as he slowly walked through the hall and the reception area and out the front door with me.

Hatfield, at seventy-two, couldn't give a damn what other people thought of him. He wanted to get back to his office. He wanted to have a drink.

I had flown to Idaho to meet this person I'd been told was a hazardous-waste-to-fertilizer pioneer reborn as "100 percent organic" fertilizer maker.

He didn't disappoint. He knew his garbage. Hatfield asked me to drive his beat-up yellow pickup truck. He smoked cigarettes as I steered the forty miles of road through the Caribou National Forest and Portneuf River valley to Soda Springs. As we drove, Hatfield told me his story. He had operated a sixteen-hundred-acre family farm in Iowa. "My dad raised fifteen-bushel-an-acre corn," he said. "I have the same farm and am raising one hundred and fifty bushels because of fertilizer."

But he didn't want to work on a farm. Hatfield left it to a tenant manager. He became a salesman, producer, and hustler—a pioneer in fertilizer.

"I've been in the fertilizer business longer than anyone because I started in fifty-three and anybody else my age is damn sure smarter and retired," Hatfield said.

The old man had spent half of his life on the margins of the industry—turning toxic waste into fertilizer. He knew all about waste from bronze mills, brass mills, galvanizing industries, and steel smelters.

He knew all the people who dumped toxic waste through the fertilizer loophole, and he knew how much money they made doing it.

He was one of them.

"Nucor Steel didn't want to ship their lead zinc dust to Monterrey, Mexico, at one hundred dollars a ton, and so they got Frit Industries to move in there," Hatfield growled. "You say how do I know that? Because they asked me to do it before Frit. Nucor worked on me over a year trying to get me to do it, but I got scared and I wouldn't do it.

"What turned me off is I talked to one of my friends there and he says, 'John, I'm only allowed to work here for six months and then I have to be moved somewhere else because of lead intake.' And I thought, 'Holy smoke, I don't want to die.'

"So I wouldn't do it. I told them to find somebody else."

Nucor got Alabama-based Frit Industries to build a fertilizer plant next door to the Nucor Steel furnace in Norfolk, Nebraska, *right* next door, connected by pipes. That way, Hatfield explained, they avoided paperwork and the safety requirements of the federal toxics law. Frit set up shop in the heart of corn country and convinced Midwest farmers that corn needed zinc, though some scientists would dispute that. And Frit sold zinc from hazardous waste at half the price of its cleaner competitors.

As we drove, I asked Hatfield how the money worked. "When I used to be in the know down there, anything over six thousand tons, Nucor was paying Frit thirty dollars a ton to take it. Because otherwise they were paying one hundred dollars a ton to ship the stuff down to Monterrey, Mexico."

Zinc Nacionale operated a recycling plant in Monterrey, 150 miles south of Texas. The high-temperature process purified zinc, cadmium, and lead from American hazardous waste. Companies like Frit, on the other hand, mixed the waste with acid and passed the toxics through fertilizer.

"It's garbage," Hatfield said. "It's a mess. It's worse than it's ever been."

We drove through an ancient seabed, the heart of the western United States phosphate industry. Hatfield—like scores of other small and midsize fertilizer makers—had built his business in the shadows of heavy industry. He picked the perfect place to be a small fish feeding off the whales.

Over six million tons of rock was extracted every year from open pit mines, most within one hundred miles of Soda Springs, the mountain town where Hatfield set up shop. Idaho produced fifty tons of 99.9 percent pure phosphorus a year—and four times that much waste.

Monsanto Company and FMC Corporation owned the giant smelters in Pocatello. They were the last two remaining elemental phosphorus plants in the United States. Monsanto used half its supply for Roundup, the top-selling agricultural chemical in the world, a herbicide that kills by hyperstimulating growth. FMC sold to Pepsi and other food companies. Phosphoric acid gives the tangy taste to soda pop.

For a while, Hatfield took the ash left over from the FMC smelting process. It was laced with toxic chemicals. Now he took waste rock from a Simplot mine.

Hatfield showed me his fertilizer factory, black piles and mud and dust everywhere, conveyor belts and roaring machinery. Hatfield bought almost two hundred thousand tons a year from Simplot at the nominal price of ten dollars a ton. His factory crushed the unwanted ore and rolled it with lignite to make granules of phosphate fertilizer. He turned Simplot waste into Hatfield profits with a brilliant marketing angle.

"I've gone organic!" Hatfield barked. "Great business to be in."

And it was true. He was certified organic. The slogan on his baseball cap said:

SODA SPRINGS PHOSPHATE
BLACK GOLD
ORGANIC FERTILIZER

"We've had a lot of fun," Hatfield growled. "I think definitely some people have thought I was a wild man. And maybe that's true. But I'm probably the most successful independent in the western United States that is producing fertilizer. There's not many of us left."

I left Hatfield's factory and drove to meet the man in charge of Monsanto's waste. Robert Geddes, Monsanto environmental manager, was also chairman of the agriculture committee of the Idaho State Senate. He was, of course, Republican. Geddes explained the recycling end of Monsanto's eight-billion-dollar operation.

"It's all about money," he said. "Everybody's trying to work together to get as much value out of this stuff as we can."

The largest waste stream was once used as filler material for road-ways and cement buildings. Now a lot of the people who lived in those buildings were unhappy about the slag mixed in cement because it turned out to be low-level radioactive. Two citizen groups were suing.

So now the slag was useless, Geddes said. It was dumped in an ever-higher mountain of waste near the Monsanto plant, where nobody would sue them later.

Monsanto's mountain is a legitimate tourist attraction, like a vol-cano. Set back two hundred yards from a chain-link fence on the bor-der of the company property, the mountain grew. Every ten minutes or so, a tractorlike device pushed a six-hundred-cubic-foot vat of molten metal up to the edge of the plateau. The truck tipped the vat over the edge, like pouring a bucket off a roof.

The molten metal crashed down the side of the mountain, break-ing roaring tumbling hissing rocks, a spectacular sight and sound, night or day. It had snowed in Soda Springs. The slag coated the side of the mountain in layers of hot lava frosting. The snow on the edges flashed to steam. Within a minute, the red surface of calcium silicate rock turned to orange and then brown. It glowed beneath the cooling crust.

Ten minutes later another giant vat was poured to the side of the first one, then another and another, lava layers back and forth on the edge of the slag plateau creeping toward the distant property line. Monsanto had operated the phosphorus refinery twenty-four hours a day since opening day in 1952. Monsanto Mountain was seventy acres and 140 feet tall. It grew by a million tons a year.

"We could sell it," Geddes said. "People ask us for it all the time. It's good filler material. But we do not sell slag anymore. We will not sell it."

Another part of the waste stream was a moneymaker. Geddes took me to the top of the plateau to show me the ferrophosphorus. It was drained out the bottom of the vats of slag and cooled to solid form, five times heavier than normal rock, like a meteorite. It was rich in iron, vanadium, and toxic heavy metals. The ferrophos was loaded in trucks and hauled two miles to a plant built in 1968 by Kerr-McGee Corporation.

The price was tied to the market value of vanadium, a metal strengthener, extracted by Kerr-McGee and sold to companies like Boeing. What remained after that was a solid waste with the unwanted, uneconomical, and unsafe elements concentrated yet again, sent to a third plant, one-half mile away.

There, Evergreen Resources made fertilizer.

I drove down Monsanto Mountain and into the muddy parking lot. It was the dirtiest fertilizer maker I'd seen in more ways than one.[3] The owner of Evergreen Resources, Allan Elias, was a tall man who had moved around from New York to California, where he dodged the tax collector, to here, where he scratched out a living recycling waste into fertilizer. He had declared bankruptcy but managed to stay in the business. He was about to be indicted for poisoning a creek and a worker. He still sold eighteen tons of Nutri-Mix fertilizer every year.

When I told him who I was and what I wanted to ask about, Elias ordered me to leave. "There is no story," he said.

His black-coated little industrial plant was set up a few hundred yards from Hatfield's equally filthy-looking Soda Springs Phosphate plant. Over the years, these two fought like junkyard dogs.

Hatfield got so mad once that he punched Elias in the face. "Wish I hadn't done that," Hatfield told me with a laugh that meant he was damn proud of it. "I've been a ba-ad boy."

Another time, while drunk, Hatfield walked into Elias's house—the door was open—and challenged the younger man to a fight. Elias pulled out a pistol and shot it into the television to scare Hatfield away. Hatfield said he'd bring his own gun next time.

These were the people to whom Monsanto, number 187 on the Fortune 500 list of America's largest companies, and FMC, number 355, gave their wastes to make into fertilizer.

I left Elias's parking lot and drove back up Monsanto Mountain. There Geddes showed me the worst waste of all. It was called electrostatic precipitator dust: ash from the smokestack. They made six thousand tons a year.

Forklifts carried heaping cauldrons of the ash, red-hot and smoking, to a corner of the property where the cinders cooled and crusted over. They were left in a growing hill of black material on top of the giant slag plateau.

Monsanto used to pay scavenger fish like Elias and Hatfield to take the ash and turn it into fertilizer. The ash was packed with zinc and laced with toxic heavy metals. For Monsanto it was a cheaper disposal than landfills. But three years before the day I stood on Monsanto's mountain, the company started piling up the ash instead of sending it away.

That was something special. Monsanto became the first company I'd discovered to in effect admit it had been taking risks for years with a toxic waste material recycled into fertilizer—then stop because of possible effects on the food chain.

"What really is a concern is product liability," Geddes explained. "Is somebody going to sue Monsanto down in time because we allowed it to be made as a fertilizer?"

Monsanto stopped the practice out of caution even though it was legal and other companies continued with EPA blessing. "Sometimes we pay for mistakes the federal government even helps us make," Geddes said.

A metals broker in Seattle offered to charge Monsanto thirteen dollars a ton, plus transportation, to send the ash to China. "It would cost us seventy dollars a ton in all. He thought if we wanted to get rid of it bad enough, we'd pay him for it," Geddes said.

Monsanto declined the broker's offer. The company didn't know what they'd do with the ash in China.

I dropped by the Idaho Department of Environmental Quality on the way to the airport. Gordon Brown knew what I wanted to know. He was the man who kept an eye on Monsanto and the fertilizer companies. He spoke so quietly, I wondered if he was afraid of being spied on in his own office.

"The industry has a lot riding on saying this recycling is beneficial," Brown told me. "If you start looking at this, you'll open a can of worms."

He spoke even softer. "You have to be very, very careful. These industries are very powerful. They have lobbyists who can get the federal government to change the rules to suit them. And if you get anything wrong on anything you say, they'll try to crucify you for that."

QUINCY, WASHINGTON

Easing into a chair in the town clerk's office, Patty faced three men who embodied forces against her: Greg Richardson, the potato grower; Pete Romano, owner of Quincy Farm Chemicals; and Dwight Gottschalk, food plant manager.

It was three on one. If this were basketball, Patty would fake up, let them fly by, then lay the ball in the hoop. But this was a face-off in a small office, not a basketball gym.

Patty used the same move. She held back, felt her confidence grow, then shot. She'd never been able to dunk the ball, but this was just as good.

It was this: She told them it was *legal* to dispose of toxic waste through fertilizer.

The men seemed surprised. *Legal?*

Yes.

Patty didn't want to be the target of a bullet from anybody who thought what they were doing was illegal.

The men knew nothing of the code of federal regulations saying you can put toxic waste in fertilizer legally. They just wanted Patty to disavow "Lead in Your French Fries?"

Romano pressed the attack.

You're going to kill the economy *here. Why are you doing this? What do you get out of it?*

Patty set her jaw. She was going to listen, not talk.

Romano rose. He stood over Patty in the chair. He was a big man. The chemical-company owner looked down at the mayor.

You're so smug, he said, going red in the face. *You're so damn smug.*

Patty thought he was going to hit her. But she didn't wince. She didn't cry. Never in public.

She looked up at Romano for a long moment.

Then he sat back down.

Later, Dwight told Patty he didn't agree with the tactic of physical intimidation. Patty said she learned something.

"As long as you can control your emotions, you can control the situation," she said. "And men do not like that. Men do not like that."

At the end of the meeting, Patty asked the silver-haired Richardson to write down exactly what he wanted from her.

To: Mayor Patty Martin
From: Greg Richardson
Subject: Notes from 4/30/97 Meeting

To address the public's concerns and fears that have been generated by the fertilizer issue, I would suggest that you consider the following in a public statement to the press:

1. *Remove yourself from any connection with the "Lead in your french fries" article and the people represented in the article.*

2. *Remove yourself from any connection and affiliation with the small group who are trying to use the situation as publicity for their own personal interests or gain (i.e. Ring Leader)*

3. *A statement of your trust in the appropriate government agencies and their ability to deal with both the Cenex situation and the waste in fertilizer issue.*

4. *A statement in regard to the safety of the food product produced in our highly regulated agricultural industry.*

5. *Assure the public that you will have no further involvement in the waste in the fertilizer issue.*

Thank you for your consideration of these items of concern.

But Mayor Martin did not trust the agencies to take care of the fertilizer or food safety. All the agencies did was tell her they wouldn't do anything.

Patty phoned me four times in the two days I was on the road in Idaho, each time leaving a bolder and bolder voice-mail message. My pep talk had helped. She would not relinquish her constitutional rights. She would not be buffaloed. She would not stop questioning.

She phoned the city attorney to ask why she shouldn't talk about hazardous wastes in fertilizer. What was the legal reason? After all, as mayor, public health was her business. The city attorney didn't phone back.

Patty had promised not to talk about hazardous waste in fertilizer *as mayor*. She wished she hadn't made that promise: She couldn't abide by it; she would try to work around it. She'd call herself Patty Martin, not Mayor Martin, and do her troublemaking from home, not the town hall.

Glenn was worried they'd be sued.

You promised you'd be quiet, he told her.

But Patty told her husband they had to take that chance.

As she dwelled on what had happened, she began to talk with Glenn about pulling the kids out of school. As a mother, their health was her first concern. The kids went to school a block from Cenex. Maybe they should pack up and leave Quincy.

"We've got four children, and if there's something happening here, we have to think about it."

She called Dennis, Tom, Duke, me.

Patty Martin was out of the bottle again, and she was in fighting form.

One day Glenn got the idea to have a laboratory analyze the dust from their furnace filter. It would show what chemicals were floating in the air in the Martin house. For less than a hundred dollars, a lab could check for ten heavy metals.

What should we look for? he wondered.

What was in Eric's hair? Patty said.

———

The Far West Fertilizer and Agrichemical Association set up a public forum on the subjects of plant food and toxic waste. Their best experts gathered a few blocks from Patty Martin's house.

Patty thought of her father whenever she entered the double glass doors of the Quincy Community Center, a former supermarket. The family had held Al Naigle's memorial service there after he had died of cancer. The place had been packed. Now half the chairs were empty, but still the talk on fertilizer attracted a hundred people despite it being a sunny workday evening in May.

Tom, Nancy, and the organic farmer Bill Weiss sat together. I sat to the side, the out-of-town reporter, specifically invited by Far West Fertilizer. The subtexts of the meeting, by now obvious, were to question Patty's questions and stop my newspaper story. Too late, I thought.

And Professor Mortvedt was there. I was sorry to see him brought to remote Quincy by the fertilizer group. It eroded his credibility as an expert. His presence said industry could deliver him. Mortvedt talked about the ubiquity of heavy metals in nature and the alkalinity needed to keep them out of plants and food.

Then he said, "You'll find that very little research has been done on heavy metals in micronutrient fertilizers." Exactly the point, I thought.

Larry Schaapman sat by Patty Martin. They talked together quietly during the meeting. Patty didn't consider Larry an enemy. She thought he'd been used by Cenex to dump on Dennis. She told Larry about new tests from the state. The tests showed secret ingredients in fertilizer from Quincy and surrounding towns.

Can I have those? Schaapman asked.

Sure. Patty handed him a few pages of test results.

One product stood out: an all-purpose farm fertilizer from Cenex Supply in Quincy. No one who bought it could have known it had 56 parts per million cadmium, which would have made it illegal in parts of Europe and Australia, or 54 parts nickel, or 235 parts chromium, or 300 parts lead.

Who knew where any of that came from?

This was a Cenex product that would add poisons to plants. Especially leafy plants. Especially in sandy soils. Like much of the Columbia Basin.

———

A week later, the mail brought news: the report on the furnace-filter dust in the Martin house. Patty called Glenn at work. He came right home. They read it together.

ELEMENT	MARTIN HOUSEDUST (PPM)	STATEWIDE BACKGROUND[4]
Cadmium	5.8	0.2
Lead	84.9	17
Nickel	85.6	38
Mercury	10.5	0.07
Titanium	804	n/a

Everything that was high in Eric's hair was high in our house dust! Patty said.

If her house had been a factory, she told Glenn, the EPA would be pounding at the door. But the Martins had no context: What is house dust supposed to look like? They found out later that metals accumulate ten times higher than normal in a dust filter. The results that scared them were actually inconclusive and impossible to interpret.

Duke Giraud's collection case went to court on a summary motion by Quincy Farm Chemicals. Duke couldn't afford a lawyer. He represented himself and talked about property rights.

The judge frowned. Duke sounded like one of those fanatics who gave judges headaches.

"When a lawyer won't represent you, this constitutional-law stuff starts to make a lot of sense," Duke said.

But he was afraid the judge would get so mad he'd land in the county jail. When Duke talked about proof of hidden ingredients in fertilizer to be revealed soon by a Seattle newspaper, though, he saw the judge's face turn from angry to thoughtful. The judge overturned the summary judgment. Duke's case stayed alive to be argued in court another day.

Dennis's lawyer was grasping for evidence in the suit against Cenex. He asked the Fertilizer Institute for reports on heavy metals in fertilizer and was told there weren't any. He asked the Food and Drug Administration for standards on heavy metals in food and was told there weren't any. He asked the Farm Bureau for reports of crop kills. His questions were met with stony stares.

"It may be that the theory of the fertilizer industry works," the lawyer told me, "that a waste is toxic in its original form, but if we spread it out in fertilizer then we've achieved the nineteen-fifties solution: Dilution is the solution to pollution. But at the same time, consumers and the public have the right to know what's going in their food."

He thought this might not be a good case for the courts. Public knowledge might be the only way to change the system.

JUNE 1997—*SEATTLE TIMES* NEWSROOM

My exposé was almost ready. I'd found widespread use of toxic chemicals in fertilizer everywhere. I sketched out stories to put Quincy in context across America:

In Denver, a professor was fighting plans to send wastewater from a Superfund site through sewage treatment and apply the sludge as Metroglo fertilizer on a fifty-two-thousand-acre, government-owned wheat farm. The Superfund site contained industrial solvents, petroleum oils, pesticides, and radioactive materials, including plutonium, americium, radium, and strontium 90. Adrienne Anderson, professor of environmental affairs at the University of Colorado at Boulder, was the lone dissenting voice on Denver's Metro sewage agency board of directors. She called the recycling plan "a ruse to foist toxic waste onto the nation's farmlands and onto our dinner plates."

In Gore, Oklahoma, a uranium-processing plant was disposing of low-level radioactive waste by spraying it on company-owned grazing land. They used ordinary farm sprinklers to spread it. Three and a half years after the shutdown of the Sequoyah Fuels Uranium Processing facility, workers were still sprinkling wastewater from a holding pond, diluted by rain, at the rate of ten million gallons a year. John Ellis, Sequoyah Fuels president, told me the company sprayed the liquid on seventy-five acres of Bermuda grass. Hundreds of cattle grazed there. The liquid was called raffinate and was registered as a fertilizer with the Oklahoma Department of Agriculture, though the chief fertilizer regulator told me she was not aware of this. Other officials had approved the fertilizer plan in 1986. Raffinate, the main waste from a solvent used to extract uranium for nuclear-plant fuel, is slightly radioactive and contains eighteen heavy metals.

"We were screaming our heads off when all this was first happening," a local lawyer, Kathy Carter-White, told me. "But it was just like the powers that be were going forward. We just felt violated by what happened because the land will never recover."

Some people blamed the fertilizer for such mutations as a nine-legged frog and a two-nosed cow. They also said it could be a factor in some of the 124 cases of cancer and birth defects counted in families living near the plant. There was no proof. Carter-White said, "It's hard to separate out what damage came from the chimneys at Sequoyah Fuels and what was from the pallets on the ground and the groundwater and the land disposal. But the frog was found by a little boy at a country pond that was real close to where this surface application was taking place. The boy shot it and turned it over, and found it had legs sticking out all over its sternum."

In two states, I heard complaints about gypsum wallboard torn from home demolitions and spread on fields as lime. Gypsum from recycled sources adds more than calcium sulfate to the soil. It also adds asbestos, fiberglass, and heavy metals in some cases. By one estimate, a recycler would have to pay sixty dollars a ton to recycle wallboard safely but just fifteen dollars a ton to crush it and spread it around a farm. Some fields were white with it.

A shiitake mushroom grower believed hidden contaminants in recycled wallboard killed his crop. He found out a product labeled "natural gypsum" contained elevated levels of arsenic, cadmium, mercury, and selenium. He couldn't prove it killed the crop, but he learned so-called natural materials bear watching, too. "I'm all for reusing things, but they need to take better care," the grower told me.

Fertilizer companies sold recycled toxic waste in blended products for backyard gardeners and golf course owners, too. Toxicologists told me traces of poison in the retail fertilizer market would be more dangerous than in farm products because homeowners lay it on so thick and leave it out where the kids and pets can touch it.

There were, of course, no warning labels. There were no limits. And I could not find a single government agency in America that had analyzed fertilizer for the unadvertised ingredients.

So I did it myself.[5]

I bought twenty common products at home-and-garden stores and delivered them, unopened, to an accredited laboratory a few blocks from the *Seattle Times*. Frontier Geosciences analyzed fourteen toxic metals.

At least six of the twenty fertilizers had undisclosed toxic chemicals from recycled industrial waste. Three flunked the Canadian standards. Three could not be checked by standard laboratory analysis. They didn't dissolve in nitric acid. I wondered how they would dissolve in dirt. They wouldn't. Their true contents were unknowable.

Sometimes the color of the granules gave them away. Clean zinc is white. Dirty zinc is black, gray, or brown.

One of those was a white plastic jar decorated with a drawing of a garden. I bought it at the True Value hardware down the street from my home. The label said:

NuLife All Purpose Trace Elements is a mixture of the most common elements.

I asked the clerk what it was. She squinted at the fine print and said, "It's like an expensive multivitamin."

I came to treasure that cup-sized jar. NuLife All Purpose Trace

Elements was nothing more or less than a highly toxic hazardous waste captured from the pollution-control device of a steel smelter, wetted with acid and rolled into dark brown granules to look like fertilizer, and sold in my neighborhood hardware store. It even smelled of metal.

I learned a smelter somewhere paid $100 to $200 a ton to get rid of the toxic waste I was holding. I paid $4.49, plus tax, for twelve ounces. No wonder, as I walked out of the hardware store, I had a spooky feeling that I was part of a vast, unknowing network of people helping heavy industry save money by sprinkling its hazardous wastes on our land.

The label listed the plant foods boron, copper, iron, manganese, molybdenum, and zinc. It made no mention of lead, nickel, or cadmium, toxic chemicals the lab found at 2,490, 515, and 86 parts per million, respectively—among the highest levels we found. If this jar were classified as a waste instead of a product, it would have to be disposed of in a fenced landfill with double plastic lining.

NuLife was made by Pace International, Limited Partnership, of Kirkland, Washington, and distributed all over the West. A company official told me they'd bought the material in bulk from Frit Industries of Ozark, Alabama. Frit was the nation's leading purveyor of hazardous-waste fertilizer.

He said it was, of course, perfectly legal.

We found even higher levels in Ironite, a fertilizer for home and golf course use. The green-and-yellow label said:

Nothing greens like Ironite. Natural Minerals. Will not burn. Even if you apply 2 or 3 times the prescribed amount.

Ironite had almost 0.5 percent arsenic and 0.25 percent lead.[6] I called them. The owner, Heinz Brungs, explained Ironite was made from a sixty-acre, 110-foot-high mountain of mine tailings near Humboldt, Arizona. He sent me an aerial picture. He was proud. This was natural. The residue from the old silver mine was crushed, rolled into granules, and sold as fertilizer in all fifty states and overseas. Because of the nature of the minerals at that mine, Brungs said, the arsenic and lead were safely bound with other metals and would never be released in an absorbable form unless they were scorched to four thousand degrees Fahrenheit or submersed in laboratory-strength acid.

Brungs came up to Seattle and ate some in front of me to prove it was safe. He brought his lawyer. I learned that Brungs, while an amiable fellow, had a reputation for threatening to sue people who questioned his product.

Ironite was registered as a "1-0-0" fertilizer, or 1 percent nitrogen, as low a grade as possible to be called a product. The recommended application rate was so high that Ironite would add twenty-three pounds of arsenic and ten pounds of lead per acre every year it was applied.

On the brighter side, our tests of twenty home products showed that five of them were cleaner in toxic metals than the background level in the soil.[7] The concentrated, crystallized chemical products like Miracle-Gro and the organic fertilizers without phosphate were the purest plant foods I found, and those were the ones I started using in my own backyard and vegetable garden.

———

For reporters in the late 1990s, the Internet was as important as the telephone.[8] The World Wide Web was my medical librarian, agronomist, biologist, toxicologist, lawyer, regulator, telephone operator, and postal carrier. So it seemed. I used the Net, newsgroups, and E-mail every day to chase down the trail I was on. There were about one hundred million Web pages and growing. I designed one, "The Reporter's Desktop," to help my own work.

On-line, I found proof of polluters sending hazardous wastes to fertilizer makers through industrial material exchanges, where industry and waste traders hooked up. The Web brought government data from the Toxic Release Inventory and the Right-to-Know Network. It brought Medline, a free index to nine million medical journal articles from around the world.

A Web search dug up scientists at Cornell University tracing the movement of toxic elements through soil to plant roots to edible parts. They wanted to reduce the cadmium content of wheat.

Trolling for experts, I found the National Lead Information Center, and there, Dr. Janet Phoenix, who said she was shocked industry would recycle lead into fertilizer. "The less of it, the better," she said.

I found Ellis Gunderson, author of the Market Basket Survey. He said the Food and Drug Administration survey of toxic chemicals in American supermarket food had found a big drop in the amount of lead in food since 1980.[9] The change came from taking the lead out of gasoline, paint, and food-can solder. Gunderson said the levels of cadmium and arsenic in food were holding steady. They still posed a health risk.

I found Richard Loeppert, a Texas A&M University scientist who studied arsenic and helped industry recycle wastes into fertilizer more safely.[10] "A lot of the industrial additives can have major, major contaminants, so this is always a concern," he told me. "My personal opinion is there's a need to improve the (fertilizer) label and list the ingredients. I could see if you have an industrial source, the public needs to know."

My editor and I talked about visual images to explain the complexities of the issue: the image of a beleaguered mayor, a mountain of waste, a bulging warehouse, a magic silo, Tom's fertilizer tank, Dennis's kids in a cloud of dust.

We talked about how to write it responsibly. Whether toxic waste in fertilizer made people sick was not proved. But we didn't need to prove it. The question had been raised scientifically and remained unanswered. As journalists, our job was to shine light on the truth as we could see it. That truth was, at its root, an issue of risks and unknowns. The effect on health was unknown. We wrote about the risks.

Who says? Among the reputable groups who said the practice posed some risk to health were the EPA, Agri-Food Canada, European and Australian scientists, industry-financed task forces in California, and the Fertilizer Institute itself. The question was not whether there was a risk, but how much.

Oh yeah? Hazardous waste in fertilizer. Hard to believe. It was still hard to believe, but we wouldn't have to rely on personal opinions or debate semantics.

A man like Dick Camp could talk circles around the semantics of the terms "hazardous" and "waste" and "fertilizer" until you conceded everything was hazardous including pure water (hazard of drowning), and nothing was waste including the worst poisons known to man (recyclable materials), and black was white, and down was up.

In writing, I tried with every turn of the story to avoid the definitional, legalistic, semantic arguments that tortured common sense. I applied the Duck Rule: If it looks like a duck and walks like a duck and quacks like a duck, it's a duck.

That was both simple and true. So if a chemical was potentially harmful to health, it was "hazardous." If it was a by-product from another industry, it was "waste." If it was put in a product said to improve the soil, it was "fertilizer."

By now I could name names across the nation. I knew the simple facts. Polluting industries saved millions of dollars sending hazardous waste to fertilizer makers. The toxics in the waste were neither limited by law nor tested by government nor disclosed to users. Farmers and gardeners were buying products containing unknown, hidden levels of unsafe toxic chemicals. If they did not know about the risks, then they could not take precautions. Those chemicals persisted in topsoil for years. They flew in the dust, were breathed by children, were absorbed by plants, and entered our food, according to careful studies by leading scientists.

Would our report cause a food scare? We thought not. Would we be sued? We hoped not but were less certain. Thirteen states had passed laws since Alar making it easier to sue for product disparagement even if you were careful with your facts.

The beef industry had dragged Oprah Winfrey into a Texas courtroom for daring to talk about mad cow disease and hamburger. Oprah would win, but we wouldn't know that till later. The product disparagement laws were probably unconstitutional—the Supreme Court has long held that the truth is an absolute defense against damages for defamation—but those laws were still on the books. We could be sued in any of those thirteen states.

The powerful interests that did not want the public to know about toxic wastes in fertilizer would scrutinize every sentence I wrote. One little mistake and they'd pounce.

———

Patty called again and again.

She wanted to confront Cenex for moving the concrete from the old rinse pond to a special landfill. It was evidence. Now it was gone.

"My council said I couldn't pursue hazardous waste in fertilizer—but Cenex is fair game. I want a letter from the city attorney saying exactly why I can't say anything about this," she said.

She needed to decide by mid-July if she was going to run again. If she did, Greg Richardson had made it clear there would be a lot of money thrown against her. George Nutter and Dick Zimbelman were talking about running against her.

By now it was clear to Patty, as to me, that she was going to be the central character in the story I was writing.

She was scared. I was safe in Seattle. But Patty and her family would take the heat in Quincy.

I tried to shore her up. I told her these facts were all a matter of public record. The farmers had threatened to sue her; the council had told her to quit or be quiet. In attacking the mayor, they had called attention to themselves and their town. It would all be told. It had to be.

I told Patty the story needed a protagonist, and she was it. Time to stand up.

But it was a real life to her in Quincy. It was not just a story.

Glenn wasn't sure what "protagonist" meant. He looked it up in the dictionary. "The tragic hero in a Greek play; leader of a cause."

"The *tragic* hero," Patty said.

CHAPTER 8

FEAR IN THE FIELDS

WE KICKED INTO HIGH GEAR when the crosstown rival, the *Seattle Post-Intelligencer,* tried to steal our story.

One Monday in June, the *P-I,* a Hearst-owned newspaper where I had once worked, sent three reporters and three photographers to Quincy, Idaho, and Oregon. The *P-I* was gunning for a quick take based on what people told them I had dug up. I would find out later that Jaycie Giraud, impatient with my pace, tipped them off.

Andrew Schneider, a charismatic, two-time Pulitzer winner recently hired at the *P-I,* rang Patty Martin's doorbell. He wanted to interview her and the Wittes.

Patty phoned me.

Should we talk with him? she asked.

I would rather you did not, I replied. *I need more time to finish the story right.*

Okay. We won't, she said.

Patty called back an hour later.

We tried, but we couldn't do it. We just sat there and blabbered. He was so sincere, and he said he believes us. I couldn't keep my mouth shut.

I asked Patty to try to find out when Schneider planned to publish the story in the *P-I.* He told her Friday. Four days away.

Back at the *Seattle Times,* we knew we had two days to wrap it up. Journalism is a competitive business, usually to the benefit of the public. The *P-I* forced our hand. We had to go now. We worked sixteen-hour days. The story memos took shape as a newspaper series. Pages were added to our press run.

We titled the articles "Fear in the Fields—How Hazardous Wastes Become Fertilizer" and published Thursday, July 3.[1] This was the key finding:

What's happening in Washington is happening around the United States. The use of industrial toxic waste as a fertilizer ingredient is a growing national phenomenon, an investigation by The Seattle Times has found.

The Times found examples of wastes laden with heavy metals being recycled into fertilizer to be spread across crop fields.

Legally.

In Gore, Okla., a uranium-processing plant is getting rid of low-level radioactive waste by licensing it as a liquid fertilizer and spraying it over 9,000 acres of grazing land.

In Tifton County, Ga., more than 1,000 acres of peanut crops were wiped out by a brew of hazardous waste and limestone sold to unsuspecting farmers.

And in Camas, Clark County, highly corrosive, lead-laced waste from a pulp mill is hauled to Southwest Washington farms and spread over crops grown for livestock consumption.

RECYCLING SAID TO HAVE BENEFITS

Any material that has fertilizing qualities can be labeled and used as a fertilizer, even if it contains dangerous chemicals and heavy metals.

The wastes come from iron, zinc and aluminum smelting, mining, cement kilns, the burning of medical and municipal wastes, wood-product slurries and a variety of other heavy industries.

Federal and state governments encourage the practice in the name of recycling and, in fact, it has some benefits: Recycling waste as fertilizer saves companies money and conserves precious space in hazardous-waste landfills. And, mixed and handled correctly, the material can help crops grow. . . .

Among the substances found in some recycled fertilizers are cadmium, lead, arsenic, radionuclides and dioxins, at levels some scientists say may pose a threat to human health. Although the health effects are widely disputed, there is undisputed evidence the substances enter plant roots.

Just as there are no conclusive data to prove a danger, there are none to prove the safety of the practice.

In other nations, including Canada, that lack of certainty has led to strict regulation. There, the approach is to limit toxic wastes in fertilizer until the practice is proven safe. Here, the approach is to allow it until it's proven unsafe.

Although experts disagree as to whether these fertilizers are a health threat, most say further study is needed. Yet, little is under way.

And so on, for three pages of broadsheet.

I held my breath. I always do after a big article. I was turning new ground and wondered how it would look to readers. Will they understand? Love it? Hate it? Was I too soft? Was I fair enough? Did I make any mistakes? Would it cause a food scare?

When I was a much younger reporter, I used to wait nervously for the phone to ring after investigative articles. Uh-oh, somebody might be upset. Since then I've learned to take the initiative. The best thing to do after such an article is reach out to the main people in it, especially the people made to look bad, and ask them whether the story—every word—was fair, accurate, and complete. Once I struggled to make the day-after calls. Now they were natural. On the record, off the record, any way you want—be real and tell me what you think.

Of course the practice helps me be more fair, accurate, and complete in the first place. That's the idea. And it unearths errors, follow-up ideas, and most important, when successful, respect. The best advice I can ever give another journalist is to make those day-after calls and be open to what they say, and I don't mean just your friends. The day of publishing a big story is always thrilling, sometimes scary. In this case I didn't have to wait long.

A bouquet of flowers was delivered to the newsroom for me a few hours after the story hit the street. They were from Schneider at the *P-I* with a note, "Always a pleasure to read good journalism." I can't say how much that meant to me as I went home after the first day of "Fear in the Fields."

Overnight, the Associated Press and Knight Ridder wire services sent a capsule version of the story nationwide. It was published by newspapers in Chicago, Philadelphia, Houston, Boston, San Francisco, Miami, St. Louis, Sacramento, and Des Moines.

The *New York Times* ignored the story, and the *Los Angeles Times*, the leading newspaper in the nation's leading farm state, printed twelve inches on page A-6. They should have climbed all over the story. But it was a complex story. It was the Fourth of July weekend. Seattle was an isolated corner.

And despite the unchallenged evidence, it was still hard to believe industries would recycle hazardous wastes into fertilizer without limits, testing, or disclosure.

The story brought a tremendous response from people who saw it across the country. More than two hundred people phoned and e-mailed me the first three days.

From Emory University, Dr. Jeff Mahr: "As an Immunologist, I can assure you that these compounds will have a profound effect on the immune system of people who eat the food that comes out of these fields—especially when it comes out in levels low enough so that they don't kill, therefore encouraging complacency."

From an EPA investigator who asked to remain anonymous: "They're dumping a lot more toxics than we know onto farms through fertilizer. The industry is trying to stop us from enforcing hazardous waste laws. . . . They can take the toxics out and make it clean. Some companies do. Others don't. It stinks."

From Christopher Feise, the Washington State University cooperative extension service liaison with the EPA: "You have done a major public service by writing about the fertilizer and waste connection. It was a surprise to me and many of my colleagues who work in agriculture."

From Tom Wilson at the EPA: "Excellent, well documented story. One critical aspect that did not get much attention but that does concern me—the ever-growing accumulation of heavy metals in the soil with each annual fertilizer application. Heavy metals don't biodegrade or move quickly through the soils. So even 'a little bit' applied each year will inevitably accumulate to disastrous levels. Where will we farm then?"

Seattle congressman Jim McDermott, a child psychiatrist, called the recycling of hazardous wastes into fertilizer "dangerous" and even "unconscionable." Promising to explore legislation, he wrote, "The fact that there is no conclusive evidence that such contaminants are safe for use as fertilizer is alarming."

Rachel Binah, a California member of the Democratic National Committee, took a reprint of "Fear in the Fields" to a reception for Vice President Al Gore and pressed it into his hand. Binah said she asked Gore, *Did you know toxic waste can be made into fertilizer?* Gore smiled and said no, he didn't know that, but he would look into it. An aide to the vice president took Binah's name and promised to call. He never did.

Two public-interest law firms wanted to jump in the fray, but weren't sure how. After all, the practice was, apparently, legal. Bill

Bean of the Columbia Basin Institute said he might try to sue under right-to-know issues, product adulteration, hazardous-waste violations, private nuisance, or civil-rights violations by the Quincy town council against the mayor.

There were complaints about the articles, too.

The Texas scientist Loeppert didn't like how he was quoted. My editor and I tried to explain we had to reduce comments he made to one paragraph to save space. That didn't satisfy him. Loeppert had been quoted saying "It's a definite problem," without the article being clear enough that he was referring to the lack of testing and labeling, rather than the practice of recycling wastes, which he was helping industry do.

Allan Felsot, the Alar defender who'd written the fertilizer report for food processors, didn't like the fact that he was *not* quoted in the newspaper. I wasn't about to give him voice, because his report was obviously biased.

The *Times* editorial page offered a forum to Pete Fretwell of the Far West Fertilizer and Agrichemical Association to write an industry response. He began: "I'd like to be among the first to congratulate The Seattle Times on their series, 'Fear in the Fields: How hazardous wastes become fertilizer.' There were facts in the Times' story that needed to be known. Environmental scofflaws have hurt the safety record of legitimate members of the fertilizer industry."

Then he complained that the articles failed to distinguish between the scofflaws and legitimate recyclers, that they didn't mention Bay Zinc could sell legally in Canada, and that they oversimplified some complex research. Fair enough, I thought. I had had to choose and condense for readers. I had wished to include more scientific detail, but the average reader wouldn't want to plow through it all.

What I did next was unusual for a journalist.

I compiled all the scientific studies, task force minutes, and analytical reports in my files, made copies, and mailed them to a dozen people on every side of the issue. I've always been willing to share information in the public domain if asked. In this case, I took the initiative. The package proved, beyond any doubt, that hazardous wastes were being recycled through fertilizer to save industry money, that the toxic elements were absorbed by plants, and that some unknown numbers of people were facing a health risk from the practice. It was the proof. It was more than five hundred pages of material.

Patty Martin called it "the mail bomb."

I also used the Internet to send out every article as it was published to an informal mailing list of more than fifty people around the globe. This way, they could read the full versions, and I could immediately solicit their feedback and follow-up ideas.

I tried to phone Patty. For her, it was more than a scoop or a story with a new angle on fertilizer.

It was her life, changed forever.

JULY 1997 — QUINCY, WASHINGTON

Patty Martin had no answering machine, no Internet, no E-mail. When she missed telephone calls, she thought people would call again. She had mixed feelings about how much she wanted to talk, anyway. If I were sued, the newspaper would defend me, but if Patty were sued, her family might stand alone.

Glenn Martin suffered through a staff meeting at Lamb Weston. Dwight was fuming after the first article. If it was a national story, why focus on Quincy?

That night, Glenn told Patty his career was over. Every word in the article could be true, but it hurt him. Patty tried to explain that the industry leaders had made it worse on Quincy by attacking her, denying known facts, drawing attention to themselves. But Glenn was the one inside the industry circle, and it was hot there.

Most people in Quincy never saw "Fear in the Fields: How Hazardous Wastes Become Fertilizer." Quincy was outside the *Seattle Times* circulation area. Patty was left to deal with what people had *heard* about the story.

Patty tried to talk with Harriet Weber one day when they both arrived to pick up children from a babysitter. Harriet had suffered the death of her own son to a brain tumor when he was four years old in the late 1980s. Dennis had thought it might have been connected with off-grade fertilizer the family used.

Harriet, what do you think of this? Aren't you concerned? Patty asked.

You are not supposed to be talking about this, Patty. You promised you wouldn't, Harriet replied. She closed the door to her car and drove away, shutting Patty out.

Patty walked away, shaking her head.

Eleven days after "Fear in the Fields," the Quincy town council gathered again. It was the end of another hot summer day, start of another hot seat for the talky mayor. Though there had been no food scare, no discernable stigma on Quincy products, the reaction was still burbling. I was writing more follow-up stories. Quincy business leaders wanted no more.

Pete Romano, the owner of Quincy Farm Chemicals, was angry. His face turned a darker pink.

"I'd appreciate it if the council would deal with this matter once and for all. The damage is done. Thank you, Patty," he said.

Greg Richardson spoke up.

"Patty, you know we've gone through this issue. You've asked the question. Good question. The state is looking into the issue," he said. "You know, I think we are headed to labeling. [Nobody's against] analyzing of fertilizers, which is good. I think it will prove that what we're doing is safe. I don't think it will create any public fear once the answer is there.

"I know you're going to keep talking, Patty. We can't stop you.

"But could you make an effort to get the whole story out? Could you make an effort to tell you've got three farmers who made big press, but you've got thousands of farmers out there that are doing a good job? They're using fertilizers that are perfectly safe, that are having no problems. They're very successful. Can you tell our story about that?"

CBS had called Greg. He had told the network producer there was no story in Quincy: cows and horses weren't dying, and neither were people, at least not from the fertilizer.

"Leave it alone," Greg told Patty. "The state's in process. Let 'em do it. We've asked you to do that before. But you just keep going. What's the point? Where are you trying to take this thing? I don't understand that, Patty. You're doing damage there. Why?"

"Right-to-know," Patty said.

Greg himself admitted he hadn't known about toxic wastes in fertilizer. Patty said it's been happening at least twenty years.

"Greg, I guess I am concerned as to why you fear my involvement in that," she said.

"I fear you're not telling the whole story. The truth."

A farmer interrupted.

"I appreciate anyone who wants to get involved in anything today," he told Patty. "But I wonder what is your expertise in this field that allows you to make statements to anyone on soil science or any other information in agriculture? Are you a farmer? Do you buy fertilizer? Are you familiar with anything other than being mayor of Quincy?"

Patty said the mayor had to watch out for public health.

"Do you know more about my health than I do?" the farmer demanded.

Another man cut in. "Did you know that your car puts out carbon dioxide? And that's a poison?"

"Yes," Patty said. "Go ahead—"

"And do you still drive it?"

Bill Weber spoke up. The apple farmer, who'd dropped his threat to sue, said Patty should tend to her own yard instead of other people's business—specifically, a leaking fuel tank in her own backyard. *Gotcha!* The tank had been illegal since the Martins bought the house in 1987, Weber said.

Patty, angry and embarrassed, said she was looking into having the tank removed.

The Quincy newspaper reporter took notes. She wrote Weber's

accusation in that week's article—and made the crusading mayor look hypocritical.

But Weber hadn't mentioned how he knew about the leaking tank. The person who had sold the house to the Martins and knew the tank was leaking then was Weber's father.

MOXEE CITY, WASHINGTON

Some people said Richard Camp Jr. was breaking the law, too. Camp said one person called him a baby killer after reading "Fear in the Fields," reprinted in the local daily newspaper in Yakima. The subject came up at church services. Members of the Camps' church gathered around and supported them. He thanked them for their trust.

Camp wrote a letter to the editor: "Bay Zinc Company is proud of its excellent environmental record. Our industry is highly regulated and monitored by federal and state agencies. Bay Zinc Company complies with these detailed regulations rigorously."

The letter had a boomerang effect.

It brought out the whistle-blowers, the insiders I'd been unable to reach. Some of the Bay Zinc workers phoned me after reading their boss's letter in the local paper. They were not proud of their environmental record. They were not following the regulations. They were sick, some of them desperately sick. And they wanted to talk.

Finally I thought I knew why Camp wouldn't let me inside his factory. Maybe he didn't want me to meet his employees. Maybe he didn't want me to see the signs posted on the walls. Signs warning of the health danger of lead and cadmium. Signs you wouldn't expect in a fertilizer factory.

The workers told me this: They were supposed to wear respirators and long-sleeved overalls to protect themselves. Usually they did. Sometimes they didn't. It got hotter than hell in there.

They always showered at the end of their shifts. Even so, they left their bathtubs and bedsheets at home shadowed with the toxic dust.

"I pretty much had it on me all the time. You could get clean and scrub and scrub and scrub and go home and a couple hours later you could get the soap to turn black," one worker said. "You get filthy dirty, and you can't get the stuff off."

Camp had his workers' blood tested for lead at least twice a year. He was scrupulous that way. If the blood lead level went above forty parts per million, the worker was tested every two months. If it went above fifty parts per million, the worker stayed out of the fertilizer production area until the blood lead dropped.

"I never broke into the forties with my test. Thirty-nine once or twice. Thirty-threes. High twenties," one of them told me.

Curiously, a state toxicologist had told me earlier that the state didn't find any reports of high lead levels in Bay Zinc workers. The toxicologist said he found one fertilizer company where twenty-six workers had high levels of lead in their blood, but it wasn't Bay Zinc and he refused to name the company.[2] *Personal information,* he said. *Proprietary.*

Right.

Now I found one of the Bay Zinc workers had severe abdominal pain and a claim pending with state health authorities. Another was seeing a doctor who told him he was allergic to the zinc. Another had extremely high levels of lead in his blood but wouldn't talk with a doctor.

"It's a cesspool," he said. "It's important to let people know about this."

The employee handbook said:

Absolutely nothing may be placed in your mouth while you are in the plant, including chewing tobacco, gum, candy, cigarettes, etc. Since the only way you can get lead contamination is by eating or breathing it, it is critical that this policy be observed at all times.

I thought, What about the rest of us? What about the people who mix fertilizer at companies like Simplot? What about the people who spread it on the field? The gardeners? The children playing in the dirt? Fertilizer goes to dirt and dust. If this is so dangerous, who is giving *them* dust masks and blood tests?

Bay Zinc was required by law to report any spill of hazardous ash outside the permitted areas, but the workers just swept it up. The factory submitted to government inspections but "they are announced well in advance, and we shut the plant down and clean and clean and clean."

When the workers talked about tougher regulations on hazardous waste, they didn't worry about their jobs. Quite the opposite.

"All we talked about was how much more money Dick would make because he'd be paid more to take it," one of the workers told me. "The whole reason Bay Zinc is there is because of these steel mills. If they send something [to a landfill] it has their name on it forever and the government says you gotta clean it up. So they like to send their waste to us because, you know, we do something with it and their name is off it and it's gone. It's a smart business move for them to send it to us, and they pay handsomely, eighty thousand or ninety thousand dollars for a railcar. Basically we're like McDonald's getting paid to take the hamburger meat and then selling it. There's good money to make in that."

But this wasn't hamburger. It was toxic waste. And these toxins don't show up in symptoms like diarrhea, cramping, and fever that mark the early stages of *E. coli*, salmonella, and other bacterial food-borne illnesses known to kill as many as nine thousand Americans a year. Heavy metals do their insidious harm not in days or weeks, but in decades. Nobody can prove cause and effect. A child with neurological damage might as well blame her mother.

Steel mills weren't the only ones using Bay Zinc to dump their messes on unwitting farmers. I learned the largest tire incinerator in the world had been sending its hazardous waste ash to Bay Zinc since 1994.

That stopped after I wrote about them in the newspaper.

Exeter Energy Limited Partnership of Sterling, Connecticut, gathered thirty thousand tires a day from Maine to Manhattan. It charged five to thirty dollars a ton. Exeter burned the tires in a twenty-nine-hundred-degree boiler, making thirty-one megawatts of electricity and consuming 95 percent of the tire weight in the process. All that remained was a black chalky ash.

Exeter's manager told me he'd stored the ash in a landfill until Camp came along suggesting a fertilizer recycling plan. Camp assured him it was legal. Camp was the expert. I thought it was odd they went to such lengths. Sterling was twenty-nine hundred miles from Moxee. At Camp's urging, the company sent twelve thousand tons of ash cross-country by truck and rail to the dirty little fertilizer factory in remote Washington State. The ash contained zinc from a rubber strengthener and lead and cadmium from steel belts.

Richard Camp's workers told me how they simply added sulfuric acid to the tire ash to roll granules for LHM fertilizer.[3] LHM meant Low Heavy Metal. The granules were black. The product was cleaner than steel-mill waste, which was gray, but they both affected workers' health. Camp sold LHM to companies like J.R. Simplot that insisted on a cleaner and more soluble zinc. He sold LHM for $253 a ton, a higher price than steel-mill waste but roughly half the price of competing products without lead or cadmium.

"It's disheartening," Kipp Smallwood of Cozinco told me. "We are paying good money for clean, raw materials for zinc and then doing everything we can to make it cleaner, and we find out that other people are allowed to put hazardous wastes out there."

The farmers who used the ash saved a few dimes an acre, at most, compared with cleaner zinc. Smallwood said the real benefit was to save industry the cost of separating the toxic elements and recycling more safely.

If the toxic ash from Connecticut had been collected in one place, it would have made a pile thirteen feet high and as big as a football field. There would have been twenty-four tons of lead and half a ton of cadmium in the pile. Soon to blow in the dust.

Contrary to what Camp had claimed, Exeter's ash and Bay Zinc's fertilizer both failed the federal toxicity test required for wastes that lack a special-interest exemption like the steel mills had. Connecticut officials said they would have cracked down sooner if they had realized Bay Zinc was a fertilizer company. The Washington State Department of Ecology ordered Camp to cease, desist, and dump 188 tons of unsold fertilizer into a hazardous-waste landfill. "We should have known about this," said Brian Dick of the ecology department. "I think some people have grabbed onto a little bit of gray in the regulations."

Camp fought back. In his own way, he was as relentless as Patty Martin, and a whole lot smoother. He said he hadn't understood the rules, and he said the ash wasn't really bad. It was just a little bit over the line for what the federal government called hazardous. "It makes great recycling. It makes a wonderful fertilizer. But because of half a part per million, we can't use it?"

In the view of most scientists, mere traces of lead and cadmium are sufficient to pose threats to some people's health, especially those who already exceed the safe limit, especially children. The World Health Organization says millions of people exceed the safe body limit of cadmium, and the General Accounting Office says one out of every twelve low-income children has high levels of lead in their blood, despite efforts to reduce it.[4]

Stephen Artus, general manager of the nation's second-largest tire-burning plant, in Modesto, California, had no idea how the other companies could get away with the fertilizer disposal scheme. Modesto Energy handled the California tire ash as a hazardous waste, paying Zinc Nacionale $181 a ton to purify it in a high-temperature process that separated heavy metals safely.

When I compared that cost with the $73 a ton Exeter Energy paid Camp to take its tire ash, I could see the allure of illegal recycling. And when I researched the law further, I learned something new about the consequences of grabbing the gray:

Once something is declared a hazardous waste, you cannot make it unhazardous just by diluting it. Anything made from a hazardous waste is considered a hazardous waste. In other words, the general fertilizers into which Camp's zinc was mixed at companies like Simplot were also, by the letter of the law, illegal. A strict enforcement of the law would have tracked and seized every blended product.

That, of course, never happened.

Camp sold hazardous-waste zinc for blending in ten states—California, Colorado, Idaho, Kansas, Nebraska, North Dakota, Montana, Oregon, Washington, and Wyoming—and in Canada, Mexico, and Australia.

AUGUST 1997—PROVIDENCE, RHODE ISLAND

The door was open so I walked in.

There was big Dick Camp from Bay Zinc, with the smooth fertilizer lobbyist P. Whitney Yelverton from Washington, D.C., at his side, and the kindly-looking scientist John Mortvedt across the table, and two dozen others. Captains of fertilizer, sitting around a table in a conference room at a hotel, suddenly uncomfortable.

People looked over, then looked away.

I was uninvited. I was crashing the Fertilizer Institute's Heavy Metals Task Force, a little-known group they wanted to keep little known. These were the people who were trafficking in toxic chemicals in fertilizer.

I'd been tipped off to the meeting and was testing: Would they really let the food-eating public know about their business?

I sat on a chair by the wall with my notebook and pen in hand, the reporter's semaphore to show this meeting was now on the record. People exchanged looks around the table. Camp went pink. Yelverton turned stone silent and a hush fell over the room. The last thing most of these fertilizer men wanted was to talk about their heavy metals in public. And soon the meeting ended and they walked out of the room.

I felt like a hound dog at a skunk party.

Every year, state regulators gather with their good friends from industry for a weekend at a three-star hotel like this. The fifty-first annual meeting of the Association of American Plant Food Control Officials was an expense-account working weekend for sixty-five regulators from thirty-eight states and Canada—and a place for sixty-three businesspeople to talk and listen, wine and dine, ingratiate and influence.

Here, the national policy on fertilizer was set.

Joel Padmore of North Carolina, the affable incoming president of the regulators group, told me later, "We don't have near the resources of the industry we regulate, so we try to operate by consensus."

Needless to say, they had never invited an investigative reporter to watch their consensus-building process.

Rufus Chaney, the veteran federal scientist and industry consultant, once told me why industry prefers state-by-state regulation. "That way, fertilizer companies control the laws that are written about them. That's a problem. As long as we have that problem, the public interest isn't going to be served."

RECEPTION COMPLIMENTS OF AGWAY, INC.; ALLIED-SIGNAL, INC.; CENEX/LAND O'LAKES; C.F. INDUSTRIES; FARMLAND HYDRO, L.P.; FRIT INDUSTRIES, INC.; GROW-MARK, INC.; IMC KALIUM; IMC VIGORO; LEBANON CHEMICAL CORP.; MISSISSIPPI CHEMICAL CORP.; PCS PHOSPHATE SALES; PURSELL INDUSTRIES; J.R. SIMPLOT; TERRA NITROGEN; THE ANDERSONS, INC.; THE ESPOMA CO.; THE FERTILIZER INSTITUTE; ZIPP INDUSTRIES.

It was a big sign by the bar. I was surprised the industry was so open about freebies for public officials. It violated many states' conflict-of-interest laws. The regulated industry paid for free drinks and hors d'oeuvres before dinner. The regulators bellied up to the bar.

Then the hotel staff loaded a thirty-six-foot table with meats, vegetables, breads, fruits, cakes, and pies. Sallow state bureaucrats don't
often see such feasts. They also enjoyed two open bars in the dining
room, passing, as they entered, another sign:

BANQUET COMPLIMENTS OF

A.H. HOFFMAN, INC.

AMERICAN CYANIDE CO.

BAYER CORP.

MONSANTO CO.

ROCHE VITAMINS, INC.

VIGORO INDUSTRIES, INC.

The next morning, a former president of the group took what was,
for them, a bold stand. Texas State Chemist George Latimer Jr., said he
wanted industry to put everything on the labels. Not just the advertised
nutrients. Everything.

"The time has come when [all] ingredients have to be listed," he
said. "It doesn't matter what industry thinks—there is going to be a
label statement."

Dick Camp rose to speak. I had to admire the guy's originality. He
didn't object to anybody knowing what was in his product, of course
not. But he said there was one problem: There wasn't enough room on
the label to list all the ingredients people don't know about.

"There are only so many square inches we can print things on,"
Camp said. "If we get down to ridiculously low levels of parts per million heavy metals, there will not be enough space to print all the things
we would have to."

Others chimed in. A lobbyist for home-and-garden giant Scotts Company told the regulators a label showing the hidden toxics would scare
and confuse consumers without giving them good information on which
to base a decision. Mortvedt said he'd been working on this for twenty-
five years. He thought the products ought to be labeled better to let the
buyers know. "I've just been wondering when this would finally occur."

They ended up, as government officials are wont to do, appointing a subcommittee to study the matter further. Latimer was named chairman. He said it would take two years, minimum, for a new rule to go through the association process.

Dick Camp got himself named a member of the subcommittee.

DONALDSVILLE, LOUISIANA

Camp's little company had its magic silo. IMC Global, one of the Big Three of fertilizer, had a magic valve. I learned about the valve from a Louisiana state inspector named Christopher Simms whose work was changed by reading "Fear in the Fields."

Simms had inspected IMC for years. He'd never thought to ask a simple question:

Are you guys using any waste in your fertilizer?

Why, yes, he was told. IMC was taking used battery acid. *Isn't that good recycling?*

Simms tracked the acid back to the battery-crushing company. It sold lead to a smelter, plastics to a recycler, and sulfuric acid to IMC Global for fertilizer. The acid was contaminated with heavy metals.

In 1997, Simms calculated, IMC used its fertilizer to dispose of 58 pounds of arsenic, 103 pounds of cadmium, 125 pounds of lead, and 127 pounds of chromium for the battery recycler. In fertilizer. Legally. And nobody knew.

Once again the toxics in industrial waste going to fertilizer had disappeared in a puff of regulatory magic. Simms learned that the used battery acid was reported as a hazardous waste while it was in the storage tank at the recycler, in the trucks that transported it to the fertilizer company, and in the storage tank at the fertilizer company. But when somebody turned a valve to send the acid into the fertilizer production area—presto!—it was no longer considered a hazardous waste.

"Once it entered the pipe, it was exempt from the law. It was no longer considered a waste, so it couldn't be a *hazardous* waste. It went to the recycling process as a product for fertilizer," Sims explained.

"Well, I like to see toxic waste recycled into roadbed or some kind of cement. But don't go putting it in my food. That's where I feed my kid."

So IMC stopped the practice. The battery acid was less than 1 percent of IMC's sulfuric acid and easy to replace with a cleaner source. But it wasn't the only company doing it. Sims was told almost every battery recycler in America sent used acid to fertilizer makers. Nobody knew how much.

People at the top of IMC Global were concerned, too. IMC mailed a survey to all its raw-material providers. It asked for complete laboratory tests of chemical traces. The orders came from Robert Van Patten, president of IMC AgriBusiness, after reading the article. "We're trying to set the pace in the industry for being strict on the amount of heavy metals in the products that we will buy," an IMC manager said.

IMC was in an ideal position to effect change from within. It was the world's biggest phosphate producer and owned the Vigoro brand name. The company operated more than 250 retail centers in the Corn Belt.

When I phoned her, Patty was happy with the news. She said, "This is the first company that really understands."

But things at home weren't so good. Patty sounded sad and far away. Glenn was still worried about a 1994 setback at work. He'd been passed over for promotion by somebody with less experience and no college degree. They both knew what that meant. Glenn's career was capped because of Patty.

QUINCY, WASHINGTON

Patty Martin shied from view. Small reactions, slights, remarks, let her know how she was viewed in her hometown. I read Henrik Ibsen's *Enemy of the People*. Patty lived it.

"I'm getting bruised pretty bad over here," she told me one day.

Greg Richardson, the potato farmer who'd denounced Patty at every public meeting, was named Grant County farmer of the year.

Gary Chandler, the orchardist who chaired the House Agriculture Committee and was featured in an article, "Chandler Defends Fertilizer Industry," was named honorary farmer of the year.

Chandler came to Quincy to have lunch with a group of community leaders. Mayor Martin was not invited. Chandler remembered Alar. He knew people who'd been ruined. Chandler didn't like the word "fear" and he did not like Patty Martin. "If I feared everything, I wouldn't eat a thing nor would I drink a thing because I don't know what science is going to come out with tomorrow and tell me is bad for me," he said.

NBC-TV aired a two-minute segment focused on Quincy. The principal of Quincy Junior High School tried to block the network photographer's view. Tom and Dennis went on camera to give the farmer's view. The fertilizer industry refused to put anyone on, then complained the segment was unfair.

As it was about to air, the Fertilizer Institute and the Agricultural Retailers Association put out a national media advisory claiming less than 1 percent of fertilizers contained hazardous wastes. They were using an absurdly narrow definition of the terms "hazardous" and "waste" and "fertilizer." Regulators and industry alike had agreed it was, in fact, a growing national phenomenon of unmeasured dimensions. If the industry wanted to ignore that, I thought they did so at their peril.

The NBC story echoed "Fear in the Fields." A lot of Quincy people blamed Patty Martin for the unwanted publicity, no matter that the predicted food scare, once again, failed to materialize.

A woman stood up at a council meeting and asked Patty when her term ended.

"December thirty-first."

"Good!"

"I'll tell you what," Patty replied combatively. "There's a whole lot of legislation to support between now and then."

Glenn's boss got angrier every time the media talked about fertilizer regulators, conferences, companies, *anything* that mentioned the genesis of the issue in the personage of the stubborn mayor of Quincy, Washington.

Letters to the editor of the local weekly attacked Patty with impunity.

A former supporter of Patty's named Bryan Westover wrote a letter to the editor saying Patty wasn't the person they thought they'd elected. He said her and the farmers' complaints went too far. He said she should resign.

That one really hurt. Westover could not be dismissed. He held respect in town, in church, and in the Martin home. Patty had a lot of sleepless nights.

"But they can say what they want. To me, it's all about the public's right to know. And until we started working on this, nobody knew about it. They wouldn't have, either. I believe that with all my heart."

—❦—

Patty was undecided whether to run for reelection. Glenn didn't want her to. But it was a decision she had to make on her own.

She thought she had suffered enough abuse, but she did not want to leave under duress. Patty waited till the last hour of the last day to decide.

Her detractors were right about one thing: Patty Martin was stubborn. She filed for reelection. This was a woman who crashed the basket for rebounds after misses. She had no quit.

She was challenged by George Nutter, the council member and State Patrol technician who'd often attacked her in council meetings, and Dick Zimbelman, the farmer, contractor, and former councilman, once a Patty pal but now keeping his distance.

Patty thought she'd lose. "I can handle that," she said, "but I won't stop asking questions."

The house across the street from the Martins' was owned by a Cenex employee who stuck an oversized plywood sign in the front yard: ZIMBELMAN FOR MAYOR. The house next door was owned by a Quincy Farm Chemicals employee with a big yard sign for Nutter.

"Patty is a very good, sincere person, but in this issue she is totally off base," the Nutter supporter told me. "When a person tells me I am poisoning the food supply, I am poisoning the land, I am very upset by that allegation. I help grow a very safe food supply, the safest in the world. I would not poison the land. I am proud of what I do, and I will continue to use the fertilizers that I sell."

Down the street, retired nurse Marjorie Thomason posted a Martin sign in her front yard. Thomason had known Martin's family since Patty was a girl. "She's always been involved in things. She comes from a family with high ideals."

Thomason said her own son farmed eight hundred acres in the nearby town of George, Washington, and was surprised by the disclosures about toxic waste in fertilizer.

"I think with some of the bigger farmers, they're getting worried," she said. "It all comes down to the dollar. Especially with food crops. 'Don't rock the boat.' That's what's worried some of them, that somebody's shaking the boat. They're jumping to conclusions. There's so much bad talk going around. They haven't read enough about it. I really believe she's onto something. That's why I'm supporting her, because I'd like to see her carry through with what she's doing. I think there's more to it than meets the eye, and it's not just Quincy."

Zimbelman said Patty should have spoken against toxics in fertilizer as a private citizen, not as mayor of a farming town. "When she speaks as the mayor, then she represents the whole community. I campaigned for Patty last time. But Patty got a little headstrong on us."

The *Quincy Valley Post-Register* covered the toxic issue mostly through the views of Martin's detractors. The editor was president of a business club. One letter to the editor spoke of a "rumor" that Martin had collected $600,000 in campaign money from environmental groups.

"Unbelievable!" she said. Patty spent just $150 from friends and $400 of her own money on the campaign.

Nutter was the best-financed candidate. His campaign advertisement listed 164 supporters, many from industry. Pete Romano, owner of Quincy Farm Chemicals, published a commentary in the newspaper and reprinted hundreds of copies to hand out at town events.

It was titled, "Sincere Acts Can Harm."

I don't question the sincerity of Mayor Patty Martin or her followers. I know they believe strongly in their fight against recycling industrial by-products into fertilizer. They perceive it as a possible health hazard to our town.

But I do question their lack of facts and science. I do doubt their efforts have made anything safer today than it was a year ago. I do believe their campaign has backfired, hurting Quincy's small town togetherness—our sense of being a true community—without helping anybody's health.

Those people she had tried to help were of little help to her now. Most of the water group lived out of town. "Poor Patty's taking the heat for all of us," Jaycie Giraud said. "The rest of us have just been kind of sitting back in the corner."

Meanwhile, employees of Cenex and Quincy Farm Chemicals canvassed door-to-door against Martin. Some of them used company radios to campaign on company time. She had to be stopped.

PRIMARY ELECTION RESULTS
QUINCY MAYOR

DICK ZIMBELMAN	290
GEORGE NUTTER	280
PATRICIA MARTIN	239

Patty wept the night she lost. She cried the next day, too. She felt the repudiation deeply.

"Well, I'm out," she told me in a listless voice on the phone. "I'm disappointed, but on the other hand, I had reservations about running again. I would like to spend more time with my family. There's a lot of appeal to staying home with my kids. I'm free to do what I want now.

"In fact, I've thought about home schooling. I don't know if I want them going to the school by the Cenex plant.

"And I'll tell you, the idea of just running away from here is very appealing."

—————

Two weeks later, Patty sent me a letter neatly handwritten in pencil.

Dwight had a lengthy talk with Glenn today about my activities, present, as well as future. Wanting to know what my agenda is, what I'm going to do next, am I running for another Public Office. He seemed very concerned that I wasn't going to drop the issue after I got out of office.

Glenn told Dwight that I had no plans, no agenda other than working to protect the health of the community. Dwight expressed disappointment that I had assured him "it would not be about Quincy" and then here is a National TV station interviewing in Quincy. Glenn told Dwight that there is a problem here and while he and Dwight only have 25 years left, what are they leaving the children? Dwight didn't understand why I wouldn't just let it drop.

They certainly weren't very intuitive. They just set me "free" to pursue this.

At the bottom of the note, she added:

It's 11:30 PM and I should be in bed, but am unable to sleep. Started writing a letter to Al Gore and will try to craft one as well to 60 Minutes. It's time to help people understand what this is all about.

Vice President Al Gore wrote back to Patty Martin with a form letter promising vigilance against pesticides.[5] The letter didn't mention heavy metals or fertilizer.

"So much for Al Gore," Patty said.

Winter in Quincy was a time to take stock and make repairs and lay plans for the season to come. Patty bought a new dress to wear to Glenn's office Christmas party. It had been a good year for french-fry sales, and the Lamb Weston crew was in a party mood. When Patty and Glenn walked into the room, both tall and dark haired, eyes turned. Glenn's boss, Dwight Gottschalk, sauntered over and gave Patty a big hug.

She thought, *Judas hug.*

All was not forgotten.

Nobody asked Patty to dance at the party. She liked to dance. She sat in a corner, making small talk, feeling outcast.

Glenn didn't know how to help his wife. He drank tequila shooters with his pals.

They left the party early. Not a word was spoken between them as Patty steered the car through snowbanked roads. She punched the button to open the garage door, eased in, and turned off the engine without so much as a sigh to relieve the tension she felt. Glenn kept a stone silence.

Many times Patty's name had come up at Lamb Weston management meetings, and Glenn hadn't known what to say then, either. He'd been with Patty for seventeen years and the company twenty years. He didn't know whether to keep going on the path he was on or try to make a change.

MARCH 1998 — WASHINGTON, D.C.

Ken Cook gathered the staff at the Connecticut Avenue headquarters of the Environmental Working Group. Cook was president of the nonprofit group. Their research, based on government reports, often earned respect in the Capitol. Intrigued by meeting Nancy Witte in Utah, alarmed by reading my articles in Seattle, Cook wanted to add to the emerging knowledge by answering one key question:

How much?

Nobody knew. The EPA couldn't say how much hazardous waste went into fertilizer. Industry played semantic games with numbers, minimizing. Cook, a soil scientist by training, understood the value of a simple fact: If you can add up how much toxic waste you're mixing with the earth, you can start to see how much you're eating from the fruits of the earth.

Six months later, the Environmental Working Group published a report, "Factory Farming—Toxic Waste and Fertilizer in the United States, 1990–1995."[6]

We found a bustling toxic commerce between factories and fertilizer makers. A total of 454 companies identified as farms and fertilizer manufacturers in the Toxics Release Inventory received 271 million pounds of toxic waste over the period 1990 to 1995. The major sources were steel mills, foundries and electronic component manufacturers. Along with nutrients like zinc and nitrogen were copious amounts of lead, cadmium and all manner of solvents and other industrial chemicals—69 different types of toxics in all. The tally for carcinogens alone came to 13.9 million pounds.

Richard Camp's company was number two on the list of fertilizer makers. Bay Zinc had turned almost two million pounds of toxic waste into fertilizer, just behind Frit Industries of the cornfields of Nebraska. The leading states sending hazardous waste to fertilizer makers were California, Nebraska, and New Jersey. And the researchers said there was far more they could not count.

Used industrial acids and ash, for instance, were exempt from EPA record keeping, though the agency knew they could carry hazardous amounts of heavy metals to the soil.[7]

The Environmental Working Group concluded that the federal

government broke its promise to track toxic wastes from cradle to grave. Cook recommended all fertilizers be tested, all labels list the toxic chemicals, and all farms that use them be warned and watched—all ideas antithetical to the industry.

"People have a right to know this kind of thing," Cook said. "Everything from the Oprah Winfrey trial to this study we're releasing today suggests we should be taking a closer look at the food supply."

This is Living on Earth. I'm Steve Curwood. A new alarm is being sounded about some of the chemicals used to grow America's food. Toxic industrial waste is routinely used to make fertilizer. It's been happening for decades, but only now is coming to light after complaints by farmers in Quincy, Washington. Their allegations have prompted a nationwide effort to examine the safety of fertilizer derived from waste. . . .

National Public Radio, May 15, 1998

OLYMPIA, WASHINGTON

Governor Gary Locke wanted to make Washington the first state in the nation to limit toxic chemicals in fertilizer and require they be listed on fertilizer ingredient labels.[8] "I think that people need to know what is going into the food supply and I think from that public education and awareness they can make their decisions, and I think the farmers themselves would appreciate that information," the governor said.

He set up a task force. Patty Martin fought her way on. Industry members tried to keep her off the task force—after all, she was only a *former* mayor, not a representative of any group—but she got the United Farm Workers to appoint her.

She joined Laurie Valeriano of the Washington Toxics Coalition, Doris Cellarius of the Sierra Club, Jon Stier of the Washington Public Interest Research Group, Greg Wingard of the Waste Action Project, and Bill Weiss representing organic farmers. On the industry's side: Greg Richardson, Quincy potato farmer and Martin antagonist, representing the American Farm Bureau Federation; lobbyists Scott McKinnie of the fertilizer association and Craig Smith of the food processors association; members of the Washington wheat growers and apple growers; and the omnipresent Richard Camp Jr.

The two sides were as far apart as peak and valley and could barely see each other in the haze. I went to six task force meetings. To make a long, long story short, I would summarize by saying: Industry won. The governor took the industry-backed plan to a legislature controlled by business. The Senate and House hearings were staged like stops on a rail line. They changed no one's direction.

Patty Martin testified once. She talked about the 150 percent increase in children's asthma and 45 percent increase in cancer deaths of children since 1980. She believed environmental toxins were the cause.

Governor Locke insisted fertilizer companies submit detailed chemical analyses. That was a first. But he accepted an industry plan to put the toxic-chemical information on the Internet, not the label. A shopper would need a computer and an Internet connection to find out what's in a bag of fertilizer. The governor insisted on standards for arsenic, cadmium, and lead. But they were looser than Canada's standards, which already allowed a doubling of heavy metals in the soil every forty-five years, and much looser than the environmental proposal of fertilizer clean as dirt. The governor promised to study dioxins but took no action. Industry lobbyists smiled while Locke signed the new law in a ceremony in his wood-paneled office.

"I'm proud that Washington will be the first state in the nation to monitor, regulate, and inform citizens about recycled materials used," the governor said.

The environmentalists said the new law was worse than nothing.

They said it legalized an unsafe practice. Patty Martin wasn't invited to the bill-signing ceremony.

"If I had been," she told me later, "I would have boycotted it."

———

The Fertilizer Institute also set up a task force to look at the recycled toxic wastes, but the institute was far from a neutral party.[9] Gene Allred of Frit Industries was a member of the board of directors, as Dick Camp had been before him: the top two hazardous-waste-to-fertilizer recyclers in America.

The task force hired a consultant to do a so-called risk assessment. The consultant wrote an article for *Farm Chemicals* magazine saying the study would establish the safety of fertilizers. He made that statement more than a year *before* the study was to be finished. The institute suggested a limit on lead that was a hundred times the existing level in most products with lead.

When the Far West Fertilizer and Agrichemical Association held a members meeting, Scott McKinnie and Dick Camp talked about how the industry "won" on the Washington law and avoided a people's initiative. They planned to carry the law to other states.

In California, Florida, Idaho, Maryland, Minnesota, Montana, New York, Oregon, and Utah, officials wrote up new rules on fertilizer toxics. Texas moved ahead by administrative fiat. But most of them bogged down. No other state passed a law even as strong as Washington's.

"There is almost no limit to this," Patty said. "Anything goes. It still goes on and on."

JUNE 1998 — OLYMPIA, WASHINGTON

The governor held a reception for environmental groups. This time, Patty Martin was invited. She drove from Seattle to Olympia with Jon

Stier, an attorney for the Washington Public Interest Research Group. He found Martin singularly single-minded.

"The whole trip," Stier recalled later, "Patty talked about fertilizer. That's all she talked about. You know the numbers on the placard of a truck carrying hazardous materials? She knows all the numbers, and the ones she doesn't know, she's got to know about. She's obsessed with this, and understandably so.

"I think Patty Martin is absolutely wonderful. Classic whistle-blower mentality—dogged, smart, persistent, sometimes down the wrong path but usually right. She took a lot on the chin. I originally thought of her as a mayor, a politician, then I realized she's not really a politician. She's just who she is. A fairly courageous and intelligent woman. This just lit a fire under her."

The same could be said of Stier, who left his job as a federal public defender to work on fertilizer after reading "Fear in the Fields."

Governor Locke and his wife, holding their newborn baby, were greeting people at the mansion door. Stier was a little nervous. He had written a letter to the editor with the headline, "Washington's Governor Falling for the Industry's Propaganda." Today Stier promised himself he wouldn't get in the governor's face. He had a carefully rehearsed greeting which he pronounced as he shook the governor's hand.

Although we couldn't agree on the fertilizer issue, I look forward to working with you in the future, Stier said.

You know, the wonkish governor replied, *it's my understanding that chicken poop has dioxin in it.*

Stier was surprised. Talk about chemical-industry propaganda: as if to say dioxins were safe for chickens. But he didn't want to argue with the governor. Stier repeated his greeting almost word for word, and walked off stiffly.

Patty was next.

Governor, she said, *this new law was a big mistake. I'll just take it that you didn't know what you were doing on this.*

The governor said it was nice to meet her. He started to turn away from Patty and look to the next person in line.

She stopped him with a hand on the shoulder. Patty held the governor's eyes and spoke in her lowest and most precise voice about health risks from the dust. One hundred percent of what you breathe is absorbed physiologically, Patty told the governor, so hazardous waste in the dust from fertilizer puts farm towns at risk.

Governor Locke said he would look into monitoring the dust in the air. He turned away, but Patty spoke up again.

I'm going to hold you to it, she said.

CHAPTER 9

LAWYERS AND LOSSES

SEPTEMBER 1998 — QUINCY, WASHINGTON

"YOU KNOW WHAT WE THINK of our former mayor, don't you?" a businessman asked me at the Quincy Rotary Club. He gave a thumbs-down sign and a mean, crooked grin.

I felt sorry for Patty. I was cocooned in Seattle; she was ostracized in Quincy. I won journalism awards; she lost friends. I could relax for a weekend; she could never give up.

A lawyer named Roger Kluck drove to Quincy one day to take the well-trod Patty Martin tour: the Cenex plant, the school, the warehouses, the fields that had been lost.[1] Kluck used to battle Horsehead Industries, a recycler and zinc maker in Pennsylvania, Tennessee, and Oklahoma. He'd moved to Seattle and wanted to help the embattled ex-mayor.

But Glenn wouldn't let him in the house. Glenn told Patty he didn't want her to meet with any attorneys in their home. He wanted nothing to do with talk of lawsuits. So Patty met the lawyer on the front porch and took him on the town tour. In the end, though, he could not help much because he could not afford to work without pay.

Patty got an Internet connection and spent part of every day exploring the World Wide Web from the computer in the laundry room connected to her kitchen. She fired off E-mails to Vice President Gore, the EPA, the U.S. Department of Agriculture, the Food and Drug Administration, the TV news show *60 Minutes*, the baby-food company Heinz.

"Something is very wrong here. They fought much too hard to shut me up, over a practice that supposedly is legal." She signed the notes, "Patricia Anne Martin, Former Mayor, City of Quincy." She heard back from none.

Patty wanted to tap the power of the organic food movement, but it eluded her. More than 275,000 people wrote the U.S. Department of Agriculture to block a rule that would have allowed organic food to be genetically modified, irradiated, or fertilized with sewage sludge.[2] Patty wondered how they'd feel if they knew about lettuce absorbing cadmium like a sponge from "all natural, certified organic" phosphate fertilizer. But few seemed to know. Organic farmers and environmentalists were still surprised when Patty told them about heavy metals in fertilizer.

Organic food sales in the United States were rising 20 percent a year in the 1990s, despite higher prices. Patty thought people would pay more for cleaner fertilizer, too, if they knew the difference. But she had to agree the pressure to clean up fertilizer was on the wane. The federal government wanted to leave it to the states, who were under industry's influence.

The dominant farm and food groups stood with the chemical companies on so-called risk assessments claiming recycled hazardous waste was safe enough to put in fertilizer. Patty said the studies didn't look long-term; didn't look at cumulative effects; didn't even start with an idea of how much of this was going on. The major environmental groups and the EPA, for the most part, sat silent. Patty thought they sold out for recycling.

She kept trying. Patty flew to Little Rock, Arkansas, for an international conference of neurotoxicologists. There she got a second chance to corner Dr. Lynn Goldman, the EPA assistant administrator for toxics, whom she'd chased up the escalator two years earlier. Patty still did most of the talking, but Goldman understood better.

"She stumbled onto something new and really made something of it," Goldman told me later. "I think she's really got something here that is important. She's very well informed, and she's been very effective. What more could you ask?"

But Goldman's answers never satisfied Patty. And Goldman resigned as assistant administrator a few weeks later.

Patty got a list of 115 people at the EPA and sent them all E-mails.

"As EPA officials," she wrote, "I am curious if it makes sense to you that the EPA base their assessment of fertilizer safety on industry-funded research? Or that fertilizer regulations be left up to the states?

"Is it wise to allow even small quantities of lead, arsenic, mercury, cadmium and dioxin [to] be spread over farmlands and gardens? Or better to spend a little more time and money to remove the harmful substances and truly recycle?

"Are there pending risks from lead, cadmium, dioxin, etc., in our food, air and soil? These are questions I can't answer, but would hope you could."

Only one replied. Bob Athman wrote Patty that "Fear in the Fields" had provoked concern, even dismay, among many EPA employees, but they were busy with other issues. "Be assured that you had many admirers among EPA staff for what you did and what you laid on the line to try to put a stop to the practice. Unfortunately, the right people at EPA did not imitate you. More pressure is required."

Patty wrote him back: "Oh, I don't intend to be quiet."

❧

Dennis got his day in court, eight years after the Cenex "product" killed his land.[3] Patty was confident Dennis would finally get the justice he deserved. She talked with him every day. She was glad he'd turned down a settlement offer so he could get a jury trial.

Michael Cooper, the judge who had tossed out Dennis's suit three years earlier only to be overruled by the court of appeals, would preside over the new trial. He set aside two weeks in Grant County Superior Court. Dennis's lawyers, Brad Jones from Seattle and Monty Hormel from Ephrata, were delighted by the judge's pretrial ruling, as a matter of law, that Cenex had engaged in nuisance and waste of Dennis's land. The jury wouldn't have to rule on the nuances of law. The jury would just decide damages.

The jury pool was selected at random from Grant County voters. When they gathered in the courtroom one sunny Monday morning, they were asked whether they or their families had business dealings with Cenex. Almost half raised their hands. Neither the lawyers nor the judge made that an issue. Hormel thought it was better to win over jurors with friendliness rather than confrontation.

Cenex lawyer Michael Tabler corralled the case inside a narrow question: Did the material that Cenex put on Dennis's property in early 1990 cause Dennis to lose his farm in 1992?

John Williams and Larry Schaapman sat at the defending counsel's table. Solid citizens, they testified without apology. Williams's federal conviction was barely mentioned. Williams admitted the Cenex material had hurt the field in 1990 and maybe in 1991, but by the next year, he said, it was healthy as virgin soil.

Dennis sat uncomfortably in the witness box. His pleading eyes looked to the jury, but he had difficulty explaining why he didn't take the land back in 1992.

"I didn't want to be liable. I thought it might be a toxic waste dump."

It wasn't easy explaining why he was suing for one and a half million dollars for mental anxiety, pain, and suffering, either. Dennis hadn't even seen a doctor. Tabler said Dennis just wanted money.

Anyone could contrast Dennis and his out-of-town lawyer with the two local business leaders. John Williams's daughter worked in the courthouse. She sat in the spectator area during part of the trial and smiled at some jurors as they walked by. Williams ended his testimony by saying he'd already suffered enough by being named in "Fear in the Fields."

The jury was out less than three hours. Dennis needed ten of twelve votes to win the civil case. The jury voted twelve to zero for Schaapman and ten to two for Cenex. The jurors said Dennis was to blame for the loss of his farm, and gave him nothing.

"Several of them thought Cenex should be punished but not by giving Dennis an award," Hormel said afterward. "It wasn't that he was incredible. They just thought he should have gone out there and farmed the land and made his payment."

Deborah Doran was one of the most influential jurors. She worked for a credit bureau that had frequent business with Cenex. Doran was aggressively outspoken against Dennis and swayed the jury with arguments that Dennis was looking for a windfall to make up for his inability to succeed as a farmer, said juror Keith Simpson. Doran said Dennis would have lost his farm no matter what Cenex did.

Tabler said jurors told him DeYoung's claim was simply preposterous. "Based on the jury instructions that the court had given them, they had to find that there was a negligent act by Cenex in 1990, but they went on to say the negligence didn't cause any damage two years later."

The Cenex lawyer said the case had no broader meaning. "If this is akin to the tobacco litigation," he asked, "why in the hell don't they sue the manufacturer of fertilizer instead of suing some little retailer?"

Cenex had sold Dennis's land in 1998. Dennis still owed Cenex $200,000 for the money the company had lent him for seed, fertilizer, and supplies. Dennis hung his head.

"It's like somebody repossessing your car and leaving you with the bill," he said. "Can they do that?"

Dennis figured Cenex was $450,000 ahead. They saved $175,000 in hazardous-waste disposal costs, made a $75,000 profit selling his farm, and billed him for the $200,000.

The judge tacked on $25,000. Dennis had to pay the Cenex lawyer.

"If there was a debtor's prison, I'd be in it," he said, cackling. "They'd ship me off to Australia by now. So I better run for Alaska before they put that back in."

Dennis's lawyer struck a deal with Cenex to forgive the debt if Dennis wouldn't file appeals. Dennis refused the deal. He got ready to file for bankruptcy, instead. That would shake the Cenex debt. He'd learned from Tom and Duke. A bankruptcy filing would also leave his lawyers without any money to cover years of expenses on the case unless they appealed and won. Hormel agreed to file a motion with Judge Cooper asking to reverse the jury verdict.

"When you're underwater, sometimes all you can do is keep clawing at the ice—and this is pretty thick ice." Dennis looked up and smiled.

Patty was bristling to file appeals, investigate the jury, phone the FBI again, and complain to judicial authorities. She'd call out the Sixth Fleet if she could.

"Patty is so militant, she'll want to fight to the last breath," Dennis said. "But I'm tired. I'm tired. I don't want to fight forever. You know, eight years—isn't that long enough? I could make sixty thousand dollars here no problem with the mill, but I don't want to stay here. Because you live in it. To end the chapter, you've got to leave."

—

Duke lost his court case, too. The Grant County judge said Duke's onions died from poor farming practices, disease, and pest infestation. The judge agreed there were trace metals, including arsenic, in the fertilizer supplied by Quincy Farm Chemicals, but no scientific evidence it caused any harm.

"It's over," Duke said. "We got blown out."

—

Tom Witte sat in the room he called, with a laugh, his office. It was a hallway cluttered with books, magazines, and an old personal computer on a table. His boys padded around the house in dirty socks. Tom kept studying, engrossed in thought, and hoping for salvation in court.

Witte got a new lawyer who volunteered his time to help correct what he viewed as an injustice. Carl Hagens, a white-haired veteran of class-action wars, told Tom, "We're where the tobacco litigation was forty years ago. If we can't stop it one way, we'll stop it another. If we can't prove consequential damages, we'll make it a consumer protection case. No right-thinking farmer would buy this fertilizer if they knew heavy metals and other contaminants were in it. If we can get this certified as a class action, the companies will have to pay the money back, and it will stop."

But the new lawyer, like the old lawyers, couldn't find an expert to prove cause and effect between heavy metals from Cenex and harm to Witte's crops and cows. And he couldn't sign up any other farmers, either, except Russ Sligar, who had nothing left to lose. Hagens said other farmers complained but refused to join the lawsuit. He said they were afraid to take on the fertilizer companies. U.S. District Judge Robert Whaley of Spokane denied the petition for class-action status.

"Now it's just me and Russ suin' them boys individually instead of a class action," Tom said. "It's not as strong a case."

Cenex countersued Witte for ninety thousand dollars in unpaid bills. Cenex took affidavits from people saying Tom Witte was a bad farmer. It was a painful time for Tom. His own neighbor said Tom's place was a mess, and in fact it was a mess, as Tom struggled to keep up with the bills and the work. His sister Nancy came down every summer to help, but he couldn't afford hired hands. Cenex field men smeared his reputation all over Grant County. If Tom's fields died and his cows died, they said, it was his own fault. Just look at him.

Terri Witte consoled Tom. He was a broke farmer, not a bad farmer, she told him one evening sitting around the chipped Formica table after the boys were in bed.

"The guys who are telling them you're a bad farmer used to sit here asking for your advice," Terri said.

"Oh, I know it's not true," Tom drawled. "You learn to toughen up to the talk after a while."

But he still swam in old debt. Tom's creditors were dogged. One lawyer in Quincy who had taken a personal dislike to Tom (and Dennis and Patty as well) chased him like a dog chases a rabbit. The lawyer worked for a former hired hand who said Tom owed him a few thousand dollars, but Tom figured he really worked for the chemical companies who wanted to get the son of a bitch who wrote "Lead in Your French Fries?" Tom had two bankruptcies going, business and personal. There was no way to pay off all his debts and back taxes and balance the books on the farm.

Tom and Nancy Witte hatched a new plan. Tom would sell the farm to Nancy and rent it back from her. He would close the business bankruptcy and wash the IRS debt through personal bankruptcy. He said, "That's the only way to save the farm."

So Tom did shift his land and equipment to Nancy's name. She owned a one-hundred-thousand-dollar house in Alaska and a one-hundred-thousand-dollar Individual Retirement Account. She made sixty thousand dollars a year from the Fairbanks school district. She was creditworthy.

So then Tom worked for his sister. He worked a hundred hours a week, and he was diagnosed with diabetes.

"It's really hard, but we've been through worse," he said. "We used to have to delay getting the water on the hay because we didn't have money. We used to have to delay hay cutting because we didn't have enough money to buy gas or diesel or bailer's twine. We used to have to buy gas by the five-gallon can. That's when you know you're really bad, when you buy it by the five-gallon can.

"Then you buy it by the fifty-gallon barrel. Then you put a couple hundred in. That's how you climb out of it. Right now we're buying gas by the five hundred gallons."

He paused. "There's not many people as good at surviving as us."

Eventually Cenex made a settlement offer to dismiss the suit over the toxic chemicals in the fertilizer tank. Witte accepted. Cenex gave

Tom $2,500 and paid his lawyer fees and wiped clean his $156,000 debt. "We settled to get it over with," Tom said.

The Internal Revenue Service bill, inflated from five thousand to sixty thousand dollars because of interest and penalties, was negotiated down to twenty-nine thousand dollars. Nancy paid half, then the IRS seized the Wittes' check from Darigold to pay the rest. "They stole the milk check," Witte said.

That made it harder for a while. Tom couldn't pay for straw for the cows. Nancy scraped up the money.

The Martin family stored some salmon in the Wittes' big freezer. One day Tom told Patty that his freezer had come unplugged and the salmon had been spoiled. He laughed nervously, as was his habit; he was sorry. When Patty told Glenn, he was angry.

He hated the Wittes. They'd attacked Lamb Weston's french fries with the title on their reckless paper, and now they ruined the fish Glenn and one of his sons had brought back from Alaska.

Glenn didn't want Patty to have anything to do with them anymore.

One evening Patty tossed a big green salad and barbequed salmon for her sister, Nancy Smith, who was a clerk at the town hall, and their families. The women started talking about the current mayor's new, mysterious illness.

Dick Zimbelman had come down with a lung disease. He'd driven over to Seattle to visit the lung specialist at the university. Dr. Raghu had told him it was idiopathic pulmonary fibrosis: of unknown origin.

Patty said it was further evidence of toxins in the dust and air.

At that, Glenn got up silently and walked away from the table.

He was fed up. Would it ever end?

Patty still watched for trucks and trains in the valley. She still believed with all her heart that Quincy was a center of a black market in toxic waste. But she could no longer get the police chief to pull over a tank truck and ask what he was hauling. And she could no longer utter more than two sentences to Dwight Gottschalk and Tony Gonzales at Lamb Weston without their anger rising like the mercury in August. Once she had counted them as friends.

She lost the support of the former regulatory attorney Kluck, too. He wrote an E-mail, "Subject: Time to fish or cut bait," insisting on a stronger voice in managing legal activity, sharing information, and controlling finances if he were to continue helping Patty try to reform the system.

"You are incredibly unfocused flitting from one thing to another on this like a butterfly," he wrote. "Patty, there are some things in this project you are very good at. But you cannot do it all. In some areas you know just enough to be dangerous. Some of your legal theories are way out there, for just one example. You need to focus on what you're good at and get the help you need in the other areas."

Kluck said Patty was making too many big promises with too little planning. "After two and a half years, I think it's time something was accomplished in this area," he scolded.

Patty sent me a copy of the letter and a one-line response with the symbol of an unhappy face: ":(What did I do to deserve this?"

About the same time, Patty was worrying out loud about some fertilizer I'd heard about which was made from a by-product of nuclear fuel processing. She appealed to Rufus Chaney for help. He called her an "extremist" who didn't understand the science.

She was just trying to raise questions. She didn't know the answers.

All she'd done was spend half a decade raising a hundred good questions in a thousand phone calls, notes, and conversations. Yes, she was scattered sometimes, and prone to paranoia, and she needed help from people who could work the system. But now she thought Kluck was looking for money. Patty never made any money off the concerns she'd raised.

Glenn Martin resented my presence. I understood some of the reasons why. As Patty had used me to bring attention to her cause, I had used her for storytelling and significance in the newspaper stories. I had used her to express the fear in the fields. She spoke from her heart. Her voice touched people. Then she suffered the retaliation, while I merely described it. Now I was safe in far-off Seattle, and Glenn was left in Quincy with the consequences.

I told Glenn his wife was a heroine. He said, "Heroine? She's stubborn—there's a difference."

Glenn wanted Patty to get a job. Without her $500-a-month mayor's pay, it was hard to make ends meet. Their house was paid for but they had four children to feed and clothe and send to college.

Dennis DeYoung wanted Marilyn to get a job, too. Patty and Marilyn lamented with each other one day, Who would you go to work for in this town? And how much abuse would you take?

The legislator Gary Chandler offered Patty a job. He told her that since she had represented the United Farm Workers on the governor's panel, and she was unemployed after losing the mayor's race, she was qualified to work in his orchard.

You can pick apples for me, Chandler said.

Patty thought that was mean and vindictive.

A filmmaker offered Patty $20,000 to sell her life rights. She said no because the contract would have allowed fictionalization. Patty worried the producers might secretly be working for the chemical industry and planning to trash her with a made-up tale.

Patty wanted to leave Quincy. Maybe go back to Alaska, the place of her happiest days. She told Glenn they should take the children away from the chemicals in the dust in Grant County. It was killing them, living there.

Glenn told her, *You're naive if you think this is just happening in Grant County. It's happening everywhere.* And he added, *What makes you think that anybody cares?*

The Wenatchee County swimming pool needed a director. Patty sent a résumé. She was a finalist but they hired another candidate. The Quincy pool needed a swim instructor. Patty offered to do it for free if her kids could swim free. The new pool director said no.

Patty felt more isolated than ever. Often she wished she'd never started down this dead end. She slumped on the couch. She stopped playing basketball on Sundays. She stopped going to church. She tried St. John's Wort, an organic antidepressant.

Many times people told the former mayor that if she had so many complaints, just leave. They weren't shy about expressing their hard feelings.

Other people, family friends, old acquaintances, looked away or hurried past on the sidewalks or in the store aisles. Patty started driving to Wenatchee, 30 miles away, to buy groceries rather than suffer the encounters in the store three blocks from her home.

The former queen of the high school sweethearts' dance couldn't even shop in Quincy any more.

EPILOGUE

IT FEELS TO ME LIKE THE STORY is just beginning. Forty months have passed since I first drove into Quincy and saw the chemical interests try to shut down Patty Martin.

She survived the shunning. One day the phone rang in the Martin house with a call for help from the man who'd been spreading fertilizer/deicer from Alcoa waste. For two years the man had told everybody including the governor's staff it was perfectly safe.

Now he'd changed his mind. In truth, he told Patty, it was unsafe, he said. And he needed her help. Nobody else would listen. The stuff was killing hundreds of acres of cropland north of Spokane, and he could prove it.

Holy cow! She was running again.

Patty has started to regain some of her footing in the community. She's had state officials and EPA officials in and out, quietly, with no action yet but ever hopeful. She remains obsessed with toxic chemicals hidden in fertilizer and poisons blowing in the dust and hazards to her children's health, and I understand why.

To her chagrin, though, I am working on other newspaper assignments. I haven't been to Quincy in a year. I think there comes a time when my own absorption in the issue gets in the way of other people who need to take responsibility.

Yet fundamental questions remain unanswered nationally and around the world. No one has seriously added up the amount of toxic acid, ash, slag, dust, and other industry waste being spread in the guise of fertilizer on the land that grows our food. No one has told the farmers, the gardeners, and consumers what they're risking in order for some polluting industries to save money. Experts say the practice is growing in America and probably everywhere. But how much? To what effect?

No one knows. That's still the kicker. No one knows.

But Patty Martin hasn't quit yet. She set up a nonprofit group called Safe Food and Fertilizer and continues pestering government officials to investigate the soil, water, and illnesses in Quincy. She still is profoundly dissatisfied with the answers to her questions.

Glenn Martin is resigned to this going on forever. He wants to stay in Quincy. With twenty-five years at Lamb Weston, he has a career to think about.

Dennis DeYoung and his family live in Alaska. They left Quincy in the summer of 1999. They are renting a big house in Ketchikan, and Marilyn is making friends with the new neighbors. Dennis got an electrician's license and a job in a state prison. He has a lot to laugh about there. The DeYoung children miss Quincy.

Tom Witte finally had a good year in 1999. The price of grain sank and the price of milk soared—bad news for wheat farmers but excellent for dairies. For the first time since the crop wreck of 1991, Witte Farms is making good money, and Tom is paying his bills on time. He's working harder than ever. Nancy comes down to help on the farm whenever she can. The Wittes are using chemical-free hay and hope to get their milk certified organic.

Duke Giraud lives in a suburb of Seattle and drives a truck for a living. He abandoned the bankrupt onion business. His wife, Jaycie, saw a story on TV one night about a city manager who'd found a great new street deicer. It was a low-grade fertilizer. He smiled. What a bargain. Jaycie called the city manager and asked if he knew what the hell he was doing. Did he know he was spreading a dangerous waste? He did not know.

Ruthann Keith keeps busy tending Appaloosas. She is saving a pile of the bad Sudan grass she bought from Cenex off Dennis's poisoned land. It is bleached white by the rain and sun and weather; no good as evidence. Ruthann saved one old, worthless horse who'd eaten the bad hay, to autopsy someday. But her chances of getting any money from Cenex fell away when Dennis lost in court.

John Hatfield lost his source of black gold. J.R. Simplot Company stopped giving Idaho mine tailings to Hatfield for a token ten dollars a ton. Simplot managers say Hatfield is no more organic than they are. The old waste recycler thinks Simplot is retaliating for him talking to me. "My big mouth," he says. Hatfield found another ore supply, but Soda Springs Phosphate fertilizer is banned in Canada and Washington State because of the cadmium.

The state issued fifty-six stop-sale orders, denied forty-five license applications, and persuaded ten companies to clean up their source materials under the new standards for heavy metals in fertilizer. One stop-sale went against a subsidiary of the seventy-five-billion-dollar German conglomerate Siemens AG for selling nuclear fuel processing waste as unlicensed fertilizer.[1]

Ironite changed its label to reduce the application rate by 90 percent. That made it a legal product despite the arsenic, cadmium, and lead bound up in the Arizona mine tailings. The new label says people should not eat Ironite. About twenty-five other products also changed their labels to comply with the new law. Same stuff, smaller doses.

Richard Camp Jr. has stopped recycling the toxic-laced steel mill waste from Oregon, brass mill waste from Illinois, and tire ash from New York and New England. Camp says he is making "a cleaner product that meets all regulations that I am aware of for fertilizer any-where." He was fined $35,000 for spilling waste with lead and cad-mium, but persuaded the state ecology and county health departments to take two-thirds of the fine in the value of Bay Zinc laboratory ana-lytical services.

Recent news reports have found asbestos in vermiculite on hard-ware store shelves[2]—enough to cause one additional cancer in every one hundred workers who use it regularly, the EPA says—and 12 per-cent cadmium in a million pounds of zinc imported from China. Fer-tilizer samples from England, Germany, Poland, Lebanon, and India also show toxic chemical traces, and we've barely begun to look.

The ultimate question is still unanswered. It is the question that gal-vanized Patty, the water group, and me:

Do toxic-laced fertilizers make my food unsafe?

Patty answers yes but lacks proof. The industry answers no but hides facts. I see the lack of knowledge as the heart of the question. *Dosis sola facit venenum:* Only the dose makes the poison. If we do not know the amount of toxics dug in, we cannot size up the danger. We still don't have our arms around it.

Mountains of toxic ash and lakes of spent acid—banned from air and water release—are tilled into soil instead. They aren't tracked, tested, or regulated as the industrial wastes they are. At some level, it is absolutely unsafe. From peanut fields of Georgia to wheat plains of Manitoba and rice paddies of Bangladesh, the metals have accumu-lated in topsoil. They don't disappear. And they've been left on garage shelves within the reach of children in boxes without warning labels.

Ken Cook wrote, "Anyone who uses fertilizer has the right to know what is in it, and whether it was made from toxic industrial waste. But beyond this basic public right to know, health officials need to know what is in the nation's fertilizer in order to protect the nation's food supply, rural communities, and farmers from toxic chemical contamination."

I think that's right. As a journalist my values start with the right to know.

My other touchstone is the principle of precautionary action.[3] It says, in short, Better safe than sorry. The principle would hold the supplier of toxic chemicals to the burden of proof showing it will do no harm. The principle would require us to examine all alternatives before diluting chemical poisons through our soil.

California is still considering a new regulation to limit arsenic, cadmium, and lead in fertilizer under Proposition 65, adopted fourteen years ago. Voters then set the product safety standard at a level that would cause no more than one in one hundred thousand people to get cancer and one in ten thousand to have a child with birth defects. Why should we accept that in plant food?

The State of California and the Fertilizer Institute published risk assessment studies saying most existing fertilizers—and much worse ones—would be clean enough to meet the standard. The EPA complains about those studies but won't take the job of regulating fertilizer safety, and neither will the U.S. Department of Agriculture. A group of state regulators, lobbied by industry, proposed some very loose standards in 2001.[4]

Risk assessment is called a science. It is really a cost-benefit analysis on the amount of sickness and death that may be allowed by an activity under study. The precautionary principle is a straightforward alternative to risk assessment. Rather than saying you can do so much harm, it says you can do no harm.

Nobody can argue that there is no harm, indeed nobody has tried. Instead the government has buried itself in toxic disregard. As I look around, I wonder, how can anyone say the same chemicals that are unsafe in the air and water would be safe in the soil, our precious farmland? The fields that grow the food my children eat are being transformed into toxic waste dumps, one season at a time.

Calling it diluted doesn't reassure me. Calling it product is doublespeak. Calling it recycling is duplicitous. We're "recycling" poisons through my family and yours.

It took Patty Martin to ring the alarm. Now we all have to answer the question: What should be done next? I think we need real oversight and testing and disclosure and change. Part of the answer may be found in nature. Why not seek to remove the toxins as our legacy to the land, or at the very least, limit them to the natural background level of the soil? Roughly a third of our fertilizers would pass that test today. What is sold as fertilizer simply ought to be cleaner than dirt.

This was a small-town story, but in one way or another, we all live in small towns on a small planet.

ACKNOWLEDGMENTS

IN WRITING A TRUE STORY, I've been careful not to merge events, change chronologies, or make up dialogue for dramatic effect. The quote marks are sacrosanct; quotations that I did not personally hear, but are accurate in substance and important for storytelling, are presented in italics. I took care that every word was true to the best of my ability as a journalist. My three years of work involved crisscrossing the country, interviewing more than three hundred people, and reading more than fifty thousand pages of documents.

I'd like to thank Elizabeth Wales, my talented and tenacious literary agent, who gave me the idea for the book by calling me out of the blue and saying, "How about a book?" I'm grateful also to my editors at HarperCollins: Trena Keating, whose insight and incisiveness frequently inspired me, and Tim Duggan, who carried the project home after Trena left HarperCollins. Thanks also to Nancy Shawn, Christy Scattarella, and Bonnie Britt for reading the manuscript.

Elizabeth Kuhn, priceless research assistant, helped me tie up many of the key points in the book, and Elizabeth Ellis, another crack researcher, barraged me with scientific and medical reports on chemicals, fertilizer, food, and health.

Thanks to Mary Keller for transcription, and to the following people who all offered information, support, encouragement, and in some cases, a paycheck: Bill Bean, Bing Bingham, David Boardman, Darlene Blair, Rufus Chaney, Dick Clever, Ken Cook, John Hatfield, Todd Hettenbach, Greg Horstmeier, Dan Jones, Ali Kashani, Shiou Kuo, Roger Kluck, George Latimer Jr., Bill Liebhardt, Alex MacLeod,

Mike McLaughlin, Eric Nalder, Deborah Nelson, Alan Rubin, Erika Schreder, Kipp Smallwood, Sherry Sontag, Greg Sorlie, David Terry, Jon Stier, Laurie Valeriano, and Bill Weiss. Special thanks to Richard Camp Jr. for giving his time to a journalist who had him in his sights. I wonder if he'll let me see the inside of his fertilizer factory now.

My thanks to the Martins, DeYoungs, Wittes, and Girauds. May the story do these people justice.

First and finally, my family inspired me: my late father, Bruce, mother, Merilynn, and brother, Scott, all fine journalists; my late brother, Terry, and older sister, Chris Wilson Fisk, and my terrific kids, Lana and Grant. Nancy Montgomery, my wife, and the best writer I know, earned my undying gratitude by her pithy suggestions, asides, and edits, and her patience and support through it all.

Duff Wilson
May 5, 2001

Notes

CHAPTER 1—SMALL-TOWN STORIES

1. Harden, Blaine. *A River Lost: The life and death of the Columbia.* Norton. 271 pp. (1996).

2. *USA v. Cenex,* CR-95-025-JLQ, CR-95-026-JBH. U.S. District Court, Yakima, WA 98901.

   U.S. Environmental Protection Agency (USEPA), Criminal Investigation Division. Case 1000-0108, Warrant 94-4153-00, USEPA, Seattle, WA 98101.

   Washington State Department of Agriculture (WSDA). Cases 157Y-91, 86Y-92 (DeYoung), 139Y-92, 9Y-93 (Giraud). WSDA, Olympia, WA 98504.

   Washington State Department of Ecology (WSDOE). *Agreed Order and Interim Action: Cenex Supply and Marketing Facility—Quincy.* WSDOE, Olympia, WA 98504.

   Washington State Department of Health (WSDOH), *Public Health Assessment, Cenex Supply and Marketing Inc.* CERCLIS #WAD058619255. WSDOH, Olympia, WA 98504.

CHAPTER 2—SUSPICIONS

1. Colorado School of Mines (CSM). *Recycling Heavy Metals from Industrial Waste.* The leading hazardous-waste recyclers in the nation attend and teach a three-day, $1,000 workshop on industry practices, including recycling to fertilizer and animal feed. The course notebook costs $595 but may be obtained for copying charges only under the Colorado public records law. CSM, Golden, CO 80401.

2. Agency for Toxic Substances and Disease Registry (ATSDR). *Toxicological Profiles for Heavy Metals.* ATSDR, Atlanta, GA 30333.

Ashton, John, and R. Laura. *The Perils of Progress: The health and environment hazards of modern technology, and what you can do about them.* London: Zed Books. 346 pp. (1998). See pp. 167–176, "Cadmium: The newest toxic link in the food chain."
USEPA. *Hazardous waste land treatment.* SW-874. Office of Water and Waste Management. Washington, D.C. (1980).
World Health Organization. *Cadmium.* ISBN 9241571349. 280 pp. (1992).

3. Farmers around the world die less from cancers, generally, and live longer than most other occupations. "Despite this favorable overall disease experience, studies often find that farmers have elevated risks for certain malignancies, including leukemia, Hodgkin's disease, non-Hodgkin's lymphoma, multiple myeloma, and cancers of the lip, stomach, skin (melanotic and nonmelanotic), prostate, brain, and connective tissue."
Blair, Aaron, and S. Zahm. *Cancer Among Farmers.* National Cancer Institute, Rockville, Maryland. Occupational Medicine 6:335–354. (1991).

CHAPTER 3—CONNECTIONS

1. Chemicals in Tom Witte's fertilizer tank:

CHEMICAL	PPM	MDL	CHEMICAL	PPM	MDL
Aluminum	800	250	Mercury	0.203	0.05
Arsenic	36	0.25	Molybdenum	N/D	100
Barium	N/D	65	Nickel	N/D	100
Cadmium	N/D	15	Selenium	N/D	0.25
Chromium	222	15	Silver	N/D	25
Copper	175	50	Titanium	42	40
Lead	217	65	Zinc	343	12.5

MDL (Minimum Detection Level)
N/D (None Detected)
Source: Brookside Laboratories, Inc.

2. Cement kilns were the third largest source of dioxin air emissions in the United States. Many kilns sent their ash to farms as soil amendments, raising the cancer risk for farm families from arsenic, beryllium, and dioxins, but in USEPA's view, an acceptable level, if quickly tilled under the soil.

USEPA. *Report to Congress on cement kiln dust. Vol 2: Methods and findings.* EPA 530-R-94-001 (1993).
USEPA. *Regulatory determination of cement kiln dust.* Fed. Reg. 60 (25) : 7366–7377 (1995).
Yake, Bill. *Cement kilns as sources of dioxin-like contaminants: An initial review focusing on Holnam, Inc., Seattle, WA.* WSDOE, Olympia, WA (1996).
Dioxins are fingerprints on many industrial wastes. Yet fertilizers are rarely tested for them. The dioxin test costs $1,000 or more, compared with $50 to $100 for a toxic-metal screening. WSDOE paid for the first test I saw of dioxins in industrial waste used as fertilizer, measured in 2, 3, 7, 8-tetrachlorodibenzo-p-dioxin toxicity equivalents. It found elevated dioxins in all five materials tested:

DIOXIN RESULTS	
Steel mill smokestack ash	815
Bay Zinc fertilizer	341
Paper mill ash for farm use	35
Bay Zinc tire ash fertilizer	5
Cement kiln dust for farm use	1

WSDOE. *State of Washington Fact Sheet: Controlling Metals and Dioxins in Fertilizer.* Pub. #98-1251. (January 1998). (www.wa.gov/ecology/pie/1998news/fert.html).
As the title above shows, the state dioxins in fertilizer described them as "controlled," rather than the opposite. Behind the scenes, officials expressed alarm and scrambled, successfully, to avoid a food scare. One result of burying the dioxin results was a lack of warning to people who could be harmed. The state toxicologist wrote in a confidential memo: "Workers who recover the dioxin-containing baghouse dusts from the foundry stacks and Bay Zinc employees who manufacture and blend fertilizer may be at risk of over-exposure to dioxin-containing dust." Laurie Valeriano of the Washington Toxics Coalition in Seattle jumped on the report. "Any one in their *right mind* knows that putting high levels of dioxin into fertilizer is a recipe for *disaster*," she said. "It's *outrageous.* Now it's time to do the *right thing* and stop them from doing this. We can no longer rely on the good will of industries. We are finally *finding out* what's going in, and on the basis of this information we have to *stop* it."

That did not happen. Pulp and paper makers, settling a lawsuit by the Environmental Defense Fund, monitored dioxins in their waste applied to land, but no other industries followed the paper industry's model. The U.S. has no standards for dioxin exposure. The USEPA Science Advisory Board is deadlocked on the issue. In 1997, Washington governor Gary Locke sought funds for dioxin studies from Agriculture secretary Dan Glickman and EPA administrator Carol Browner, but they put him off. A small group of soil scientists led by Shiou Kuo of Washington State University put dioxins in a study that is, at this writing, pending.

In June 2000, the EPA, settling a lawsuit by the Sierra Club and Washington Toxics Coalition, agreed to propose standards for dioxins in waste-derived fertilizer by April 2002. The agency also agreed to seek public comment on an outright ban on dioxins in fertilizer and a reporting system for all hazardous wastes used to manufacture soil amendments. (*The Sierra Club and Washington Toxics Coalition v. EPA and TFI*, #98-1564 and Consolidated Cases, U.S. Court of Appeals for the District of Columbia Circuit, Washington, D.C.). (www.watoxics.org/tf.htm).

See also these books on health dangers from synthetic chemicals. None talk about the risks from fertilizers, which were unimagined:

Carson, Rachel. *Silent Spring.* Houghton Mifflin. 368 pp. (1962).

Colborn, Theo, Dianne Dumanoski, and John Peterson Myers. *Our Stolen Future: Are we threatening our fertility, intelligence and survival?—A scientific detective story.* Penguin Books. 336 pp. (1997).

Commoner, Barry. *Making Peace with the Planet.* The New Press. 293 pp. (1990).

McPhee, John. *Encounters with the Archdruid.* Noonday Press. 245 pp. (1971).

Schettler, Ted (Ed.), G. Solomon, M. Valenti, and A. Huddle. *Generations at Risk: Reproductive health and the environment.* MIT Press. 417 pp. (1999).

Steingraber, Sandra. *Living Downstream: An ecologist looks at cancer and the environment.* Perseus Press. 320 pp. (1997).

Wargo, John. *Our Children's Toxic Legacy: How science and law fail to protect us from pesticides.* Yale Univ Press. 402 pp. (1998).

3. The recycling scheme by Alcoa subsidiary Northwest Alloys and L-Bar Products is detailed in Cases #91-03789-K17, U.S. Bankruptcy Court, and #CS-94-0154-JLQ, U.S. District Court, Spokane, WA; in Case #91-1356CV *(Behrman v. L-Bar)*, Circuit Court of Oregon, Washington

County, Hillsboro, OR; and the L-Bar files held by state agencies in Olympia, WA.

CHAPTER 5—LEAD IN YOUR FRENCH FRIES?

1. The Heavy Metal Task Force was studying California's Proposition 65, also known as the Safe Drinking Water and Toxic Enforcement Act of 1986, which declared as law: "No person in the course of doing business shall knowingly and intentionally expose any individual to a chemical known to the state to cause cancer or reproductive toxicity without first giving clear and reasonable warning to such individual."
California Department of Food and Agriculture (CDFA). *Minutes and Agenda Materials, Heavy Metal Task Force.* Fertilizer Inspection Advisory Board, CDFA, Sacramento, CA 95814. (1992 et seq.).
———, *Development of Risk-Based Concentrations for Arsenic, Cadmium, and Lead in Inorganic Commercial Fertilizers.* Foster Wheeler Environmental Corp., Sacramento, CA 95814. (1998).
California Department of Toxic Substances Control (CDTSC). *Enforcement Case, Chemical & Pigment Co., Pittsburg, CA.* EPA ID #CAD009159476. (1994) CDTSC, Sacramento, CA 95812.
———, Riley, Norman, memo to Rick Robison. *Comments on Draft.* CDTSC, Sacramento, CA 95812. (June 21, 1996).

2. Scientists at the Brandon, Manitoba, Research Centre of Agriculture and Agri-Food Canada were watching cadmium in wheat very carefully—and quietly. They found the cadmium content of Canadian Western Amber durum wheat was often higher than 0.1 kg^{-1} grain weight—the maximum limit under consideration by the Codex Alimentarius Commission of the Food and Agriculture Organization (FAO) of the World Health Association (WHO), which develops rules for international sales of farm products and food.
So Canada's wheat exports were at risk. The scientists worried about flaxseed and barley as well. Part of the cause was cadmium in phosphate fertilizer. The Canadian scientists found that zinc would reduce cadmium in the edible grain and stems. The zinc competes with cadmium, its neighbor in the periodic table of elements, at plant roots. Fertilizers lacking zinc were found to mobilize cadmium.
Rufus Chaney, a senior scientist for the U.S. Department of Agriculture and world expert on cadmium in the food chain, said the Manitoba stud-

ies were conducted "in absolute secrecy" for the first two years because of Canada's concerns about wheat exports. By 1995, the Canadian scientists were speaking at a global workshop on cadmium in crops.

Bailey, L. D., C. A. Grant, and R. J. Hill. *Uptake of cadmium by crop plants.* Organisation for Economic Co-operation and Development (OECD) Workshop on Fertilizers as a Source of Cadmium. Stockholm, Sweden, Oct. 15–22, 1995. pp. 174–179. OECD, Paris, France. (1996).

Choudhary, M., L. D. Bailey, and C. A. Grant. *Effect of zinc on cadmium concentration in the tissue of durum wheat.* Can. J. Plant Sci. 74:549–552 (1994).

FAO/WHO 1995 Joint Committee on Food Additives and Contaminants. *Position paper on cadmium* (prepared by France). The Hague, Netherlands. 32 pp. (1995).

Grant, C. A., W. T. Buckley, L. D. Bailey, and F. Selles. *Cadmium accumulation in crops.* Can. J. Plant Sci. 78:1–17 (1998).

McLaughlin, M. J., K. G. Tiller, and A. Hamblin. *Managing cadmium contaminations of agricultural land.* OECD Workshop on Fertilizers as a Source of Cadmium. pp. 189–217. OECD, Paris, France. (1996).

3. To the chagrin of the Fertilizer Institute, Cozinco's Web site showed test results with lead in forty-five fertilizers, including these:

PRODUCT COMPARISON TABLE

SAMPLE TYPE	TOTAL %LEAD	TOTAL %ZINC	SOLUBLE %ZINC	LOCATION SAMPLE COLLECTED
35.5% ZnSulfMono	~0	38.10	93.85	Denver, Colo.
20% Oxysulfate	2.37	28.10	.03	Aurora, Neb.
36% Oxysulfate	1.61	40.50	4.85	Missouri Valley, Iowa
21% Oxysulfate	1.92	23.50	.19	Taylor, Texas
10% Zn Chelated	.003	10.50	93.00	Uvalde, Texas
Micro Mix	.14	9.50	.12	Buffalo, Texas
31% Oxysulfate	.07	29.60	16.03	Ontario, Canada
15% Oxysulfate	1.87	26.75	3.51	Stratton, Colo.
18% Oxysulfate	2.03	14.50	5.23	Goodland, Kans.

Cozinco's own product was first and purest on the list. Kipp Smallwood explained that Cozinco fertilizer came from a relatively clean industrial waste, galvanizer skimmings, for which it paid 40 percent of the London Metal Exchange price for zinc. Cozinco added water and sulfuric acid to produce a zinc molecule most available to plants, $ZnSO_4H_2O$, and then cleaned it further. Impurities were settled, filtered, and removed in a cakey material of about 8 percent lead and 13 percent cadmium. Cozinco sent the cadmium and lead cakes to Encycle, Inc., of Corpus Christi, Texas, at a cost of $250 a ton, for final, safe recycling into separated metals.

CHAPTER 6—THE MAGIC SILO

1. Agricultural Research Service (ARS). *Agricultural Uses of Municipal, Animal and Industrial Byproducts.* Conservation Research Report #44. ARS, USDA, Beltsville, MD 20705. (1998).
   League of Women Voters. *The Garbage Primer: A handbook for citizens.* Lyons & Burford. 181 pp. (1993).

2. The Resource Conservation and Recovery Act, 42 US Code 692, says: "Wherever feasible, the generation of hazardous waste is to be reduced or eliminated as expeditiously as possible. Waste that is nevertheless generated should be treated, stored and disposed of so as to minimize the present and future threat to human health and the environment."

3. The USEPA in 1978 proposed a rule that would have set limits and required consumer labels on all commercial products that recycled hazardous wastes onto the soil. This would have exposed the toxic recycling in fertilizer long ago. But the draft rule, sent to an unspecified list of outside reviewers in 1978, was neither made into a final proposal nor adopted by USEPA. Excerpt from Draft Section 3004 regulations, Timothy Fields Jr., Office of Solid Waste, EPA, Washington, D.C., Sept. 25, 1978:

Soil conditioning products made from hazardous waste shall not contain amounts of substances that, when applied at rates recommended by the manufacturer, will cause the substance addition rates to be exceeded.

Substance Addition Rates (in kg/ha/yr):

> Arsenic 1.34
> Cadmium 0.08
> Lead 0.03

Note: Soil conditioning products made from hazardous waste may contain amounts of substances that, when applied at rates recommended by the manufacturer, will cause the substance addition rates to be exceeded, provided that the facility owner/operator applies labels to packaged products and supplies the recipients of bulk products with pamphlets or similar literature that shall: (1) state that the product is made from hazardous waste and shall not be applied to food chain crops, and,

(2) list the levels of arsenic, cadmium, and lead in mg/kg on a dry weight basis.

4. The hazardous-waste laws of the United States are contained in the U.S. Code, Title 42 ("The Public Health and Welfare"), Chapter 82 ("Solid Waste Disposal"), beginning at Section 6901. Congress set a goal of "minimizing the generation of hazardous waste and the land disposal of hazardous waste by encouraging process substitution, materials recovery, properly conducted recycling and reuse, and treatment." In short, reduce, reuse, and recycle. Congress gave broad rule-making power to the EPA. Trying to meet the needs of industry, USEPA excluded many materials from the restrictions of the law:

Code of Federal Regulations, Title 40 (Protection of Environment)
. . . Section 261.4 Exclusions
. . . (b) The following solid wastes are not hazardous wastes:
. . . (4) Fly ash waste, bottom ash waste, slag waste, and flue gas emission control waste, generated primarily from the combustion of coal or other fossil fuels . . .
. . . (7) Solid waste from the extraction, beneficiation, and processing of ores and minerals (including coal, phosphate rock and overburden from the mining of uranium ore) . . . (i) Slag from primary copper processing; (ii) Slag from primary lead processing; . . . (xii) Air pollution control dust/sludge from iron blast furnaces; . . . (xix) Chloride process waste solids from titanium tetrachloride production; (xx) Slag from primary zinc processing.
(8) Cement kiln dust waste, except . . . for facilities that burn or process hazardous waste.
. . . Part 266- Standards for the Management of Specific Hazardous Wastes and Specific Types of Hazardous Waste Management Facilities
. . . Subpart C - Recyclable Materials Used In A Manner Constituting Disposal:
. . . Sec. 266.20 Applicability.
. . . (b) Commercial fertilizers that are produced for the general public's use that contain recyclable materials also are not presently subject to regulation provided they meet these same treatment standards or prohibition levels for each recyclable material that they contain. However, zinc-containing fertilizers using hazardous waste K061 that are produced for the general public's use are not presently subject to regulation. [Appendix VII, Part 261, says K061 is "Emission control dust/sludge from the primary production of steel in electric furnaces" and contains hexavalent chromium, lead and cadmium.]

For more on USEPA activities at this time, see: Lash, Jonathan, K. Gillman, and D. Sheridan. *A Season of Spoils—The Reagan administration's attack on the environment.* Pantheon. 385 pp. (1984).

5. At least two steel mills in Oregon and one in Washington paid Bay Zinc to take their hazardous waste for fertilizer. The mills later abandoned the practice for more expensive, but safer, waste disposal. A 1997 annual report says Oregon Steel Mills "chose not to pursue this opportunity because of potential risks associated with the use of these materials as a soil supplement."

Oregon Steel was found to have violated the worker safety law by exposing its own employees to dangerous levels of arsenic, cadmium, and lead
in the same furnace dust that was shipped to Bay Zinc to turn into fertilizer. (*Oregon Occupational Safety & Health Division v. Oregon Steel
Mills*. Docket SH-95426. Citation J3801-021-95. Salem, OR 97301
[April 1997]).

6. Frit Industries, a private company based in Ozark, AL, manufactures fer-
tilizer ingredients in Walnut Ridge, AR, Norfolk, NE, and Chesapeake,
VA. The Frit plant in Arkansas is a Superfund site contaminated with haz-
ardous substances. See: Arkansas Department of Ecology Permit #451-
AR-5, EPA ID #ARD059636456, Little Rock, AR; RCRA Dockets
VI-1-82, VI-415-H, VI-517-H, CSN-38-0019, USEPA, Washington, D.C.;
Frit v. Chevron, Case #J-C-92-274, U.S. District Court for the Eastern
District of Arkansas.

7. Canada limited nine toxic chemicals in fertilizers and other soil conditioners unregulated in the United States and most other nations:

CANADA'S MAXIMUM ACCEPTABLE METAL CONCENTRATIONS

BY-PRODUCTS CONTAINING 5% N OR LESS AND
REPRESENTED FOR SALE AS FERTILIZERS OR SUPPLEMENTS

Arsenic	75 mg/kg
Cadmium	20 mg/kg
Cobalt	150 mg/kg
Mercury	5 mg/kg
Molybdenum	20 mg/kg
Nickel	180 mg/kg
Lead	500 mg/kg
Selenium	14 mg/kg
Zinc	1850/mg/kg

Trade Memorandum T-4-93 (1996), Fertilizers Act of Canada. While
those numbers may at first seem straightforward, in practice the maximum concentration in products varies widely according to the recommended application rate of the product. The lower a manufacturer's
recommended application rate, the higher the contaminants can be. The
soil loading rates were developed in 1978 by professional judgment of a
team of Canadians studying sewage sludge. The rates were applied to fertilizer in 1993, though sludge and fertilizer are chemically dissimilar. The

limit was set at a level Canadian scientists believed would not double the background level of toxic metals in the soil in less than forty-five years of normal usage.

Blair, Diane H., and L. Webster. *Metal Standards for Fertilizer and Supplement Products in Canada: Derivation of standards and background information.* Fertilizer section, Canadian Food Inspection Agency. Nepean, Ontario, Canada. (1997).

Trade Memorandum T-4-112 (1994) listed twenty-eight wastes and by-products used as fertilizers or soil amendments in Canada. They were animal slaughterhouse refuse, baghouse dust from metal and battery industries, blood meal, bone char, bone meal, cement kiln dust, coal fly ash, cocoa husks, compost, elemental sulphur, fish scrap, food waste from restaurants and canneries, galvanizing fluid, garbage tankage, gypsum from wallboard, hoof and horn meal, lignosulphonates, manure, mine tailings, municipal sewage sludge, newsprint, phosphogypsum, plant ash, pulp and paper sludge, slag waste from smelting and purifying ore, soybean meal, waste lime, and whey powder.

To me, this memo revealed the vast range of waste spread on food-growing land. Under Canadian law, those materials must be disclosed, tested, and approved for safety to food, users, and the environment. The burden of proving safety is on the industry. An offender may be sent to prison for two years and fined up to $250,000. Canada's enforcement record is unclear, however, as the country lacks a Freedom of Information Act and treats many government records, including these, as trade secrets.

For more information on cadmium studies in Europe and other parts of the world, see the reports of OECD, FAO/WHO, the United Nations Environment Programme, the agriculture offices of the European Union, and the International Lead and Zinc Study Group (London). For details of proposed and existing limits on cadmium content in fertilizer in nations of Europe, see "Phosphate Fertilizers and the Environment," Appendix B, *Proceedings of an International Workshop,* International Fertilizer Development Center, Muscle Shoals, Alabama (1992). Australia set cadmium limits at 350 parts per million in phosphorus fertilizers, or 10 parts per million in fertilizers with less than 2% phosphorus. For more information: Mike McLaughlin, Principal Research Scientist, Department of Soil and Water, the University of Adelaide, Australia.

8. John Mortvedt is a longtime soil scientist at the Tennessee Valley Authority's national fertilizer research center in Muscle Shoals, Alabama. He coedited *Micronutrients in Agriculture* (Soil Science Society of America, Madison, Wis.). Mortvedt's most pertinent studies, by date: *Uptake by wheat of cadmium and other heavy metal contaminants in phosphate fertilizers.* Journal of Environment Quality. 10:193–197. (1981).

   Plant uptake of heavy metals in zinc fertilizers made from industrial by-products. Journal of Environment Quality 14:424–427. (1985).

   Cadmium levels in soils and plants from some long-term soil fertility experiments in the United States of America. Journal of Environment Quality. 16:137–142. (1987).

   Crop response to level of water-soluble zinc in granular zinc fertilizers. Fertilizer Research, 33:249–255. (1992).

   In an unpublished paper at a symposium in 1992, Mortvedt wrote that many other industrial waste products could be used in fertilizer, if they were declassified as hazardous waste. (*Use of industrial by-products containing heavy metal contaminants in agriculture.* Minerals, Metals & Materials Society, 1992.)

9. Rufus Chaney is a research agronomist with the Environmental Chemistry Laboratory, Agricultural Research Service, U.S. Department of Agriculture, Beltsville, MD 20705. Some of his pertinent work, by date: *Cadmium transfer to humans from food crops grown in sites contaminated with cadmium and zinc.* pp. 65–70. In L. D. Fechter (ed.) *Proceedings of Fourth International Conference.* Combined Effects of Environmental Factors. Johns Hopkins University. Baltimore MD. (1991).

   Cadmium, lead, zinc, copper, and nickel in agricultural soils of the United States of America. Journal of Environment Quality. 22:335–348. (1993).

   Risk Based Standards for Arsenic, Lead and Cadmium in Urban Soils. (ISBN 3-926959-63-0) Deutsche Gesellschaft fur Chemisches Apparatewessen, Chemische Technik und Biotechnologie e. Frankfurt, Germany. 130 pp. (1994).

   Phyto-availability and bio-availability in risk assessment of Cd in agricultural environments. pp. 49–78. In OECD Proceedings Sources of Cadmium in the Environment (Stockholm, Sweden, 1995). Organisation for Economic Co-operation and Development, Paris, France. (1996).

   Sources, potential adverse effects of and remediation of agricultural soil contaminants. pp. 323–359. In Contaminants and the Soil Environment in the Australasia-Pacific Region. Kluwer Academic Pub., Dordrecht. (1996).

Soil cadmium as a threat to human health. In M. J. McLaughlin and B. R. Singh (eds.) Cadmium in Soils, Plants and the Food Chain. Kluwer Academic Publ, Dordrecht. (1998).

Comments to The Fertilizer Institute and EPA regarding inapplicability of sludge loading limits as a standard for ensuring health protection from exposure to metals in inorganic fertilizers. Washington, D.C. (1999).

10. The companies that paid Herman Parramore Jr. of SoGreen Corp. in Tift County, Georgia, to take their hazardous waste for fertilizer were Atlantic Steel, Florida Steel, Georgetown, Owens Electric Steel, and U.S. Foundry & Manufacturing. They later had to pay to help clean up the site of the fertilizer factory. It cost more than $10 million to stabilize the dust by mixing it with cement and trucking it to a lined, hazardous-waste landfill. Nobody paid damages to the peanut growers, whose ruined crops went virtually unnoticed, for which they were thankful. A Tifton community outcry centered instead on charges of environmental racism from the neighborhood where the waste was stored.

For details see: www.ban.org "Poisonous problems," by Arlie Porter, *Charleston (SC) Post and Courier,* June 12, 1992, and subsequent articles; "Bangladeshi angry over case delay" *op cit.,* Feb. 6, 1995; "U.S. Toxic Waste Sold As Fertilizer in Bangladesh," *Toxic Trade Update,* #6.1, Greenpeace (1992); "Poison Fields: Dumping toxic 'fertilizer' on Bangladeshi farmers" by Ann Leonard, *Multinational Monitor* (April 1993); "Deadly fertilizer and its future threat to the environment," by Ahsan Uddin Ahmed, Bangladesh Centre for Advanced Studies, *Grassroots,* pp. 27–35, 1993; and "No to industrial wastes," *The Dhaka Daily Star,* Bangladesh Nov. 13, 1993.

11. About 3,400 tons of hazardous waste from metal recycling industries in the United States, thick with lead and cadmium, were shipped from South Carolina to Bangladesh, applied by hand on rice fields and absorbed in the crops. Some was also sent to Australia. Gaston Copper Recycling Corp., Gaston, SC; its parent, Southwire Co., Carrollton, GA; and employees of Stoller, Southwire, and Hy-Tex Marketing, Inc., of Beaufort, SC, pleaded guilty in 1992 to violating the Toxic Substances Control Act. Stoller declared bankruptcy. More recently, Stoller resumed business.

12. For details on worldwide industry work to redefine toxic waste as non-waste when it goes to recycling, from the environmentalist viewpoint, see: "Basel Convention—Complete Analysis," *Toxic Trade Update*, #7.1, Greenpeace (1994); and "The Basel Ban: A triumph over business-as-usual" by Jim Puckett, *Basel Action Network*; Seattle, WA 98122. (Oct. 1, 1997).

CHAPTER 7—POWER AND PROOF

1. Alar, the trade name for daminozide, was a growth regulator to make apples firmer, redder, and less likely to drop off trees before harvest. It permeated the fruit and could not be washed or peeled away. CBS's *60 Minutes* broadcast " 'A' is for Apple," on Feb. 26, 1989, based on the Natural Resources Defense Council report, "Intolerable Risk: Pesticides in Our Children's Food." Uniroyal withdrew its food use licenses for daminozide within a year, and the USEPA cancelled all food use registrations within two years. (57 Fed. Reg. 46434, 46437–46440 [1992]). Apple growers, complaining that animal studies did not prove a human risk of cancer, filed a $250 million class-action suit against CBS. The U.S. Supreme Court in 1996 upheld without comment an appeals court decision dismissing the growers' lawsuit *(Auvil v. CBS 60 Minutes)*. Nevertheless, food and chemical industries and farm groups, claiming Alar was a false alarm, or "scare," persuaded thirteen states to pass laws making it easier to sue people who "disparage" agricultural products. (See "The Alar 'scare' was for real—and so is that 'veggie hate-crime' movement," *Columbia Journalism Review,* Columbia University, New York [Sept.–Oct. 1996]).
The product disparagement laws, however, have not yet been used, nor have they faced a constitutional challenge. In my experience, they nonetheless have a chilling effect, discouraging some journalists from writing about food safety issues. (Peterson, Melody. "Farmers' right to sue grows, raising debate on food safety." *New York Times*. [June 1, 1999]).

2. The professor's curriculum vitae showed research funding of $88,000 from American Cyanamid from 1993 through 1996; $298,000 from fifteen companies including Chevron, Union Oil, fertilizer makers FMC and Monsanto, and the Illinois Fertilizer & Chemical Association between 1982 and 1994; and with four colleagues, $338,000 from the Illinois Hazardous Waste Research and Information Center from 1986 through 1992. In addition to the $5,000 for his literature review on fertilizer,

Felsot obtained $47,000 from Northwest food processing companies from 1996 through 1998.

Felsot wrote that my request for information on his funding "impugns my integrity as an academic, questions my objectivity, and implies that there is something fundamentally wrong with taking grant and gift money from industry. My record shows that I have received substantial amounts of grant and gift money from the chemical industry over the last 18 years, but close inspection of the grant titles shows that these have involved research on environmental chemistry of pesticides, not health effects." Indeed. So why was he asked to write a paper summarizing research about heavy metals from fertilizer in diets? And how could he draw the conclusion that there was "little effect of fertilizer use on metal content in food"? It was far outside his specialty. Felsot held degrees in entomology, the study of insects. One of his current listed research activities was "remediation of pesticide waste in soil."

3. The Monsanto and Kerr-McGee plants near Soda Springs, ID, were contaminated Superfund sites. (Record of Decision, USEPA, Seattle, WA 98101 [1995]).

Allan Elias, who made fertilizer from Monsanto waste, was later given the longest-ever sentence in the nation for an environmental crime. A federal judge in April 2000 sentenced Elias to seventeen years in prison and $6 million restitution for sending a worker to clean a cyanide storage tank without safety gear. The twenty-year-old worker suffered permanent brain damage.

4.

BACKGROUND LEVELS IN WASHINGTON SOIL (PPM):					
Arsenic	7	Lead	17	Nickel	38
Cadmium	1	Mercury	0.07	Selenium	1
Cobalt	36	Molybdenum	2	Zinc	86

Ames, Kenneth, and E. Prych. *Background Concentrations of Metals in Soils from Selected Regions in the State of Washington.* U.S. Geological Survey. Report #95-4018. (1995).

WSDOE. *Natural Background Soil Metals Concentrations in Washington State.* Pub. #94-115. WSDOE, Olympia, WA. (1994).

The natural background level of cadmium in the state report is slightly higher than the level given in the lab report for the Martins. The level varies from 0.2 to 1.0 parts per million.

5. Experts who saw the home-fertilizer test results agreed on some advice for backyard gardeners:

- Watch out for phosphate, zinc, and iron products. They are the most likely to contain toxic metals.
- Use reputable organic products or crystallized fertilizers, which dissolve in water.
- Ask manufacturers and stores to disclose fertilizer ingredients. Ask for more information than the manufacturers disclose on the Material Safety Data Sheet. Ask for the complete chemical analysis.
- Be careful not to exceed the recommended application rate.
- Keep an eye on your children and pets. A typical child eats 0.2 grams of soil a day from putting dirty hands in his mouth. Infants and mentally retarded children may eat significantly more. Lead and dioxins are among the hazardous chemicals found in some fertilizers that pose more of a risk being eaten directly than entering the food chain in plants.

6. State officials acted when they saw our laboratory analysis. Ironite was in about 25,000 households in Washington State alone. The state responded with the toughest action yet (short of a product removal) against a company that often threatened to sue those who challenged the safety of its products.

News release
For immediate release—May 8, 1998
Home fertilizer poses possible health hazard
OLYMPIA—State health officials said today that a home fertilizer product known as Ironite could be dangerous to health.

"We are concerned about the amount of arsenic and lead in this fertilizer, especially since there are no warning labels on the package . . ."

According to Denise Laflamme, a toxicologist with the state Department of Health, the health effects of arsenic ingestion can include flu-like symptoms such as abdominal pain, diarrhea, and vomiting and even death at higher exposures.

"Depending on how much arsenic is actually absorbed into the body after it is eaten, accidental ingestion of less than ½ teaspoon of this fertilizer may be toxic to small children," she said. . . .

Heinz Brungs, owner of Ironite Products Company, flew to Seattle with his attorney to meet with state officials and me. Brungs insisted Ironite was perfectly safe because the arsenic and lead were bound up so tightly with the other metals—arsenic with arsenopyrite, lead with galena—and would never pose a risk to health or the environment. Brungs said he ate Ironite like a vitamin from time to time to restore the natural minerals in his own body.

Brungs had read "Fear in the Fields." He condemned what happened in Quincy and at Bay Zinc. Those people recycling hazardous wastes, he said, were scoundrels. On further questioning, Brungs admitted Ironite had been banned in Canada in 1996. He said that was Canada's mistake, and he said the market wasn't big enough to justify the cost of legal appeals.

The Canadian regulator, Darlene Blair, told me later she had listened to all of Brungs's arguments years ago and turned him down. She said the metals in Ironite could break down in the soil over many years, and if the metals were insoluble, maybe the nutrients were, too.

Washington officials calculated that the recommended application of Ironite mixed six inches deep could raise the arsenic in clean soil to the level of a Superfund site in one year. (Homes are exempt from Superfund cleanup.) Brungs refused to pull the product from store shelves. In the end, the state issued another consumer warning, and Brungs changed the recommended application rate on the label of Ironite to a lower level to pass the Washington standard.

7. The five products that passed the "cleaner than dirt" standard were Miracle-Gro, Peters Professional, Miracid, and two Lilly-Miller liming products. Six more products—Smith & Hawken Organic Fertilizer, Walt's Organic Fertilizer Rainy Day Blend, Black Leaf Granular Iron, NuLife Spring Feed, and Schultz Bloom-Plus—just missed.

Further information on home products:

- *Seattle Times.* "Tests of 20 home and garden fertilizers" (May 17, 1998). "State rules turn back 5% of fertilizers" (March 25, 2000). "Washington state gardeners can now check levels of toxic metals in garden products" (March 26, 2000). Articles available through a Web search by title at www.seattletimes.com.

- WSDOA fertilizer product database lets you search by product name or manufacturer on the Web at www.wa.gov/agr/pmd/fertilizers#database. The results show nutrient levels, which are on labels, and nine heavy metals, which are not. No other state has a publicly accessible database.

8. Some of the best Internet sites for my work were:

Medline www.ncbi.nlm.gov/entrez/query.fcgi?db-PubMed

Find Law www.findlaw.com

Hoover's Online www.hoovers.com

Right-to-Know Network www.rtk.net

National Agricultural Library www.nalusda.gov

Google Search www.google.com

The Reporter's Desktop (my homepage) www.reporter.org/desktop

9. The Food and Drug Administration's Total Diet Study is covered in J. Assoc. Off. Anal. Chem. Int'l 71:1200–1209 (1988) for 1982–84 results, 78:910–921 (1995) for 1984–86 results, and 78:1353–1363 (1995) for 1986–91 results. For a discussion of the FDA study, see D. L. MacIntosh et al, *Dietary exposures to selected metals and pesticides,* Environ. Health Perspect., 104 (2) :202–209 (1996).

10. Loeppert analyzed twenty-four soil products, finding the toxic chemicals highest in rock phosphate, followed by sewage sludge and phosphorus fertilizer. Toxic elements were lowest in organic soil amendments and liming materials, potassium fertilizers and, purest of all, nitrogen fertilizers. Loeppert began his report: "The continuous application of large amounts of fertilizers and other soil amendments to agricultural land has raised concern regarding the possible accumulation of toxic levels of their trace element constituents and potential harm to the environment. Furthermore, increasing amounts of urban and industrial wastes, which may contain significant quantities of heavy metals, are being disposed on the land."

Raven, K. P. and R. H. Loeppert. *Trace element composition of fertilizers and soil amendments.* J. Environ. Qual. 26:551–557 (1997).

Loeppert wrote, "Fertilizers and soil amendments can contain significant amounts of potentially hazardous trace elements of geologic or man-made origin. The possibility of soil and environmental pollution through the application of these materials to agricultural lands has therefore raised some concern."

Raven, K. P., J. W. Reynolds, and R. H. Loeppert. *Trace element analyses of fertilizers and soil amendments by axial-view inductively-couples plasma atomic emission spectrophotometry.* Commun. Soil Sci. Plant Anal., 28:237–257 (1997).

His repeated use of the word "concern" led me to believe Loeppert was in fact concerned about the practice. After publication of "Fear in the Fields," however, Loeppert wrote a letter saying his remarks had been taken out of context. He said there was *industry and scientific* concern but no need for *public* concern. Loeppert said the chance of any one person being hurt by fertilizer was very, very small, and that industry acted responsibly. Loeppert said he helped about five companies each year decide whether to recycle waste materials into fertilizer. He said my article caused some companies to "hunker down" and stop asking for advice. "There's fear in the fertilizer industry, too."

CHAPTER 8—FEAR IN THE FIELDS

1. "Fear in the Fields: How hazardous wastes become fertilizer" *Seattle Times*, PO Box 70, Seattle, WA 98101 (July 3–4, 1997). Available on the Internet at www.seattletimes.com/news/special/#fields.
   Credible follow-up articles in other media included:
   "EPA, state officials looking at labeling metals in fertilizer to prevent contamination," Bureau of National Affairs (BNA) *Daily Environment Report*, Washington, D.C. 20037 (Aug. 18, 1997).
   "A New U.S. Waste Strategy Emerges," *Rachel's Environment & Health Weekly*, Annapolis, MD 21403 (Aug. 21 and 28, 1997), beginning: "A new strategy for disposal of hazardous materials is emerging in the U.S. After years of unsuccessful efforts to gain public acceptance of waste disposal in the oceans, in landfills, and in incinerators, frustrated environmental officials at the federal and state levels now advocate spreading hazardous materials onto and into the land, essentially dispersing dangerous toxins into the environment, leaving no fingerprints." On the Web at www.monitor.net/rachel/rehw-home.html.
   "Toxic wastes in fertilizers may be poisoning American food and ruining fertile farmland all in the name of recycling heavy metals," NBC Nightly News (Oct. 15, 1997).
   "Controversy over toxic wastes in fertilizer hits 'prime time': lawsuits threatened," *Green Markets*, Bethesda, MD 20814 (Oct. 20, 1997).

"Heavy Metal Fertilizer" and "Zinc's problems are many," Greg
Horstmeier, *Farm Journal*, Philadelphia, PA 19102 (January 1998).
"Industry dominates fertilizer panel," (Moscow, Idaho, and Pullman,
Wash.) *Daily News* (Jan. 31, 1998).
"Field of Battle," *(Eugene, Ore.) Register-Guard* (Feb. 1, 1998).
"Study shows K061 may be worthless," George Anthan, *Des Moines
Sunday Register* (April 19, 1998).
"Concerns about Toxic Fertilizer," Terry FitzPatrick, *Living on Earth*,
National Public Radio (May 15, 1998).
"EPA rules let heavy industries sell toxic wastes to fertilizer companies,"
Wall Street Journal (March 27, 1998).
"Fertilizing or Contaminating?" *Environmental Health Perspectives*,
Research Triangle Park, NC 27709 (March 1999).

2. Steve Whittaker said twenty-six fertilizer workers had blood lead
levels greater than 25 ug/dl at a place he refused to identify. (Whittaker
memo to Greg Sorlie, Washington Department of Labor and Industries
[Sept. 11, 1997]). He wrote of three situations where fertilizer workers
could be exposed to heavy metals: manufacturing or blending; applying
to fields; and contacting the soil after application. The U.S. Public Health
Service set a goal for the year 2000 to eliminate all worker exposures
leading to blood lead levels greater than 25 ug/dl of whole blood.
(SHARP Technical Report 38-5-1997. Labor and Industries, Olympia,
WA 98504. [1997]).

3. Bay Zinc LHM ("Low Heavy Metal") fertilizer made from tire ash con-
tained up to 1,900 parts per million lead and 48 parts cadmium, dry
weight. The toxic leaching test showed up to 5.2 parts lead and 1.1 parts
cadmium, exceeding the federal toxicity limit of 5 and 1, respectively.
Bay Zinc Co. v. EPA, Case #98-1279, D.C. Circuit Court, Washington,
D.C. (1998).
Bunnell, Ross. *Memo to Robin Bray*. Department of Environmental
Protection, Hartford, CT 06106. (Oct. 3, 1997).
Seattle Times. "Toxic ash from tires used in fertilizer." (Nov. 23, 1997).
USEPA. *Phase IV Land Disposal Restriction Implications for 'Waste-
Derived' Fertilizers*. RCRA Dockets F-97-2P4P-FFFF, F-98-PH4S-FFFF.
USEPA, Washington, D.C. (1997).

4. The General Accounting Office of the U.S. Congress reported in 1998:
"Despite the decline in the incidence of high blood levels of lead in recent
years, high concentrations of lead, causing learning difficulties and/or

severe health problems, continues to be the number one environmental hazard for young children." On the Web at www.gao.gov.

5. The USEPA sent a form letter of its own to citizens and members of Congress who asked the Clinton administration to respond to "Fear in the Fields." It said: "EPA shares your concern that fertilizers be safe and recognizes that the presence of heavy metals and other potentially toxic compounds in these products deserves greater scrutiny. . . . EPA will identify whether appropriate government actions are needed. . . ."

The EPA attached a "Fact Sheet." Later, I saw a preliminary draft of the "Fact Sheet." One section had been cut. This was the section USEPA cut:

What problems can be caused by fertilizers containing toxic materials?

Fertilizers containing toxic materials not related to agricultural needs can cause problems in a variety of ways. For example, these toxics may directly affect crops, causing poor growth, or outright crop loss. In other instances, toxics may need to build up in soils over many years before they reach levels harmful to crops. Toxics may also be absorbed by plants, and become part of the food chain when eaten by animals or humans. Fertilizers containing toxics can be harmful directly to humans when dust from fertilizers or wind-blown soils is breathed by farmers or nearby residents.

Later still, the USEPA formed an internal group to study toxic waste in fertilizer. But the USEPA had a conflicting role as recycling promoter and police. It was busy, too, with pesticides, dioxins, endocrine disruptors, and other concerns. The EPA said it would need to expand its charter to take in fertilizer, and every time Congress started looking at the EPA charter, bad things happened. Fertilizer was perhaps the last major commodity unregulated by the federal government. That was okay with the EPA, and industry liked it that way, too.

USEPA did a risk assessment on the heavy metals arsenic, cadmium, and lead in fertilizer, a study best described by one organizer as "quick and dirty." The agency took the narrowest possible view of the terms "hazardous," "waste," and "fertilizer" in determining that the practice was, as far as it could tell, safe and beneficial. The USEPA review failed to explore loopholes for dioxins, used acids, ashes, and soil amendments not classified as fertilizer.

USEPA. *Estimating risk from contaminants contained in agricultural fertilizers.* (1999) Prepared by Research Triangle Institute for Office of Solid Waste, USEPA, Washington, D.C., 20460.

USEPA. *Background Report on Fertilizer Use, Contaminants and Regulations.* EPA 747-R-98-003. (Jan. 1999) Prepared by Battelle Memorial Institute for Office of Pollution Prevention and Toxics, USEPA, Washington, D.C. 20460.

6. The Environmental Working Group report, "Factory Farming—Toxic Waste and Fertilizer in the United States, 1990–1995," is the best study yet of the scope of the practice. Excerpts here by permission:

INDUSTRIES SENDING TOXIC WASTE TO FARMS AND FERTILIZER COMPANIES

RANK	STANDARD INDUSTRIAL CODE	LB.	%
1	Steel Works, Blast Furnaces, and Rolling and Finishing Mills	79,932,179	30%
2	Electronic Components And Accessories	52,812,315	20%
3	Industrial Organic Chemicals	23,538,608	9%
4	Coating, Engraving, And Allied Services	21,690,344	8%
5	Secondary Smelting And Refining Of Nonferrous Metals	20,261,853	8%
6	Rolling, Drawing, And Extruding Of Nonferrous Metals	19,444,463	7%
7	Industrial Inorganic Chemicals	7,915,093	3%
8	Soap, Detergents, And Cleaning Preparations;	7,653,790	3%
9	Miscellaneous Fabricated Metal Products	5,226,688	2%
10	Primary Smelting And Refining Of Nonferrous Metals	5,200,000	2%
11	Meat Products	4,698,921	2%
12	Metal Forgings And Stampings	3,097,307	1%
13	Agricultural Chemicals	2,293,156	1%
14	Nonferrous Foundries (castings)	2,226,004	1%
15	Petroleum Refining	1,727,987	1%
	All others	13,375,209	

COMPANIES SHIPPING TOXIC WASTE TO FARMS AND FERTILIZER COMPANIES

RANK	COMPANY	CITY	STATE	LB.
1	Nucor Steel	Norfolk	NE	26,219,034
2	Atlantic Steel Ind.	Cartersville	GA	17,570,000
3	Allco Chemical Corp.	Galena	KS	12,700,750
4	Cascade Steel Rolling Mills	McMinnville	OR	12,597,492
5	Hoechst-Celanese Chemical	Pasadena	TX	9,191,044
6	Oregon Steel Mills Inc.	Portland	OR	8,244,876

RANK	COMPANY	CITY	STATE	LB.
7	Schuylkill Metals Corp.	Baton Rouge	LA	7,900,000
8	Armco Inc.	Sharon	PA	7,534,950
9	Photocircuits Corp.	Glen Cove	NY	6,764,632
10	H. Kramer & Co.	Chicago	IL	6,427,575
11	Magnesium Corp. of America	Rowley	UT	5,200,000
12	Seidel Inc.	Waterbury	CT	4,558,796
13	Stepan Co.	Elwood	IL	3,590,000
14	Nicca USA Inc.	Fountain Inn	SC	3,589,790
15	Phelps Dodge Magnet Wire	El Paso	TX	3,552,361
16	Phibro-Tech Inc.	Union City	CA	3,476,824
17	Bloom 'n' Egg Farm	Bloomfield	NE	3,466,400
18	Metalplate Galvanizing Inc.	Birmingham	AL	3,182,950
19	Metalplate Galvanizing Inc.	Atlanta	GA	3,111,825
20	Gulf Coast Recycling Inc.	Tampa	FL	3,107,716
21	Zycon Corp.	Santa Clara	CA	3,092,642
22	Midstates Wire	Crawfordsville	IN	3,071,509
23	Continental Circuits	Phoenix	AZ	2,936,918
24	PQ Corp.	Kansas City	KS	2,738,352
25	Harvard Ind. Inc.	Spencerville	OH	2,622,686
26	Nucor Steel	Darlington	SC	2,581,156
27	Philson Inc.	Watertown	CT	2,167,908
28	Ilco Unican Corp.	Rocky Mount	NC	2,054,552
29	Hadco Corp.	Derry	NH	2,022,000
30	Hadco Corp.	Owego	NY	1,972,000
31	Mueller Brass Co.	Port Huron	MI	1,889,966
32	Macklanburg Duncan Co.	Gainesville	GA	1,834,616
33	Cyprus Rod	Chicago	IL	1,817,842
34	Herco Tech. Corp.	San Diego	CA	1,796,855
35	Metalplate Galvanizing Inc.	Birmingham	AL	1,607,850
	All others			84,900,050

FERTILIZER COMPANIES RECEIVING TOXIC HEAVY METALS

The top five took 76 percent of the toxic heavy metals.

RANK	COMPANY	CITY	STATE	LB.	%
1	Frit Industries	Norfolk	NE	2,189,481	27%
2	Bay Zinc	Moxee	WA	1,897,556	24%
3	Tri Chem	Atlanta	GA	970,000	12%
4	Hynite Corp.	Oak Creek	WI	595,523	7%
5	Stoller Chemical Co.	Jericho	SC	462,782	6%

TOXIC CHEMICALS RECEIVED BY FARM AND FERTILIZER COMPANIES

RANK	CHEMICAL	LB.
1	Zinc And Zinc Compounds	90,374,599
2	Copper And Copper Compounds	48,820,033
3	Sulfuric Acid	34,590,979
4	Ammonia	25,348,640
5	Phosphoric Acid	17,647,789
6	Ammonium Nitrate (solution)	13,014,399
7	Hydrochloric Acid	6,850,444
8	Diethyl Sulfate	6,317,400
9	Lead And Lead Compounds	6,210,260
10	Manganese And Manganese Compounds	5,322,546

STATE RANKINGS—TOP TEN

Toxic waste shipped from these states:

Received by farms and fertilizer companies in these states:

RANK		LB.	RANK		LB.
1	California	37,677,18	1	Nebraska	30,099,831
2	Nebraska	36,869,24	2	California	29,941,974
3	New Jersey	29,733,15	3	Oregon	25,862,573
4	Washington	20,863,52	4	Georgia	23,692,539
5	Georgia	18,850,27	5	Texas	16,706,742
6	Kansas	15,539,13	6	Kansas	16,392,667
7	Virginia	14,755,40	7	Illinois	13,988,540
8	Texas	14,657,04	8	New York	10,387,105
9	Indiana	9,474,89	9	Louisiana	8,873,327
10	South Carolina	8,864,45	10	Pennsylvania	8,825,078

Methodology: EWG obtained USEPA Toxic Release Inventory records on hazardous-waste transfers between 1990 and 1995, highlighted the transfers to fertilizer or farming businesses by Standard Industrial Code (SIC), and made telephone calls to double-check where possible. The SIC listed fertilizer companies, but did not show how much of the toxic waste they received actually ended up in fertilizer.

"Factory Farming" listed three legal loopholes for toxics to flow into fertilizer: (1) the K061 exemption that allowed steel companies to send their smokestack ash with no tests; (2) an interpretation of the hazardous-waste law that permitted its use as fertilizer if the material was considered safe for landfills lined with double plastic; and (3) "landfarming," a euphemism for the practice of allowing companies to transfer wastes directly to non-food-growing farms if they can be safety rendered harmless in the opinion of company scientists.

The Fertilizer Institute said the EWG grossly overestimated the amount of toxic waste actually going to fertilizer or farm fields. The researchers at EWG stood by their findings, saying there were so many loopholes that they certainly underestimated the amount substantially. The Toxics Release Inventory itself missed many toxic wastes, including dioxins, acids, and the slags, drosses, and ashes from the copper, brass, bronze, steel, and galvanizing companies that the USEPA said in 1997 should have been counted as toxic wastes all along.

Savitz, Jacqueline D., Todd Hettenbach, and Richard Wiles. *Factory Farming—Toxic Waste and Fertilizer in the United States, 1990–1995.* Environmental Working Group, 1718 Connecticut Ave., N.W., Washington, D.C. 20009 (March 1998) and on the Web at www.ewg.org. *Wall Street Journal.* "EPA rules let heavy industries sell toxic wastes to fertilizer companies." March 27, 1998.

7. USEPA allows unknown quantities of used industrial acids and ash to be exempt from hazardous-waste reporting, though the agency has known since the 1970s that they may pose a hazard in fertilizers and soil conditioners. See: Weiner, Lawrence. *Actual and Potential Soil Amendments Made from Industrial Waste.* Memo to Tim Fields, Program Manager. USEPA, Washington, D.C. (1979).

The hazards were known to include lead and mercury from waste sulfuric acid, heavy metals from smelting acids, heavy metals and radionuclides from fly ash and flue gas residue. See: SCS Engineers. *Land Cultivation of Industrial Wastes and Municipal Solid Wastes: State-of-the-Art Study.* Volume II. Contract #68-03-2435, USEPA, Washington, D.C. (August 1978).

No one knows the quantity of used acids substituted for cleaner acids. An Alabama farmer told me his 1993 cucumber crop was killed by contaminants from used acids in fertilizer. "Some loads are completely clear and some loads are real muddy," Barton Willoughby said.

Industrial ash is spread on farmlands in the millions of tons. Nearly 600 power plants in the United States alone produce over 100 million tons of ash, sludge, and boiler slag from fossil fuels each year, laced with dioxins, furans, and heavy metals including arsenic, cadmium, chromium and mercury. Incredibly, nobody knows where all this goes, but a good share is spread on the land as lime substitute or soil amendment. Congress exempted the bottom ash, slag, and scrubber sludge from USEPA authority in 1980 by the so-called Bevill Amendment (Pub. L. 96-482). See: USEPA. *Report to Congress: Wastes from the combustion of fossil fuels.* EPA 530-R-99-010. Washington, D.C. (1999).

In March 2000, three environmental groups asked the USEPA to designate as "hazardous" the combustion wastes from burning coal and oil, requiring better tracking and disposal. The USEPA denied the request. See: *Laid to Waste: The dirty secret of combustion waste from America's power plants.* Citizens Coal Council, Hoosier Environmental Council, Clean Air Task Force, Washington, D.C. (March 2000). On the Web at www.cleanair.net.

8. Because of the governor's interest, Washington State developed the best information on toxics in fertilizer in the world. The conclusions of state reports are cautious and conservative in order to protect farm and export interests. But the data is clear: Among a wealth of information, Washington State identified the types of fertilizer with the most contaminants from both natural and industrial sources:

FERTILIZERS MOST LIKELY TO CONTAIN TOXIC METALS	
FERTILIZER MATERIAL	METALS OF CONCERN
Rock phosphates	Cadmium, Selenium
Ferrous sulfates	Nickel, Lead
Zinc sulfates	Lead
Wood ash	Arsenic, Lead
Processed leather meal	Mercury

Washington State Departments of Ecology and Agriculture. *Report to the Legislature: Levels of Nonnutritive Substances in Fertilizers.* As required by RCW 15.54.433. Olympia, WA. (December 1999).

———. *Screening Survey for Metals in Fertilizers and Industrial By-Product Fertilizers in Washington State.* WSDOE Pub. #97-341. (December 1997).

———. *Final Report: Screening Survey for Metals and Dioxins in Fertilizer Products and Soils in Washington State.* WSDOE Pub. #99-309. (April 1999).

———. *Dioxins in Washington State Agricultural Soils.* WSDOE Pub. #99-333. (November 1999).

9. The industry group lacks credibility on hazardous wastes. Its pattern of conduct shows the Fertilizer Institute (TFI) is covering up more than opening up. TFI cultivates clout in state and federal agencies. TFI internal memos show the institute planned to "manage the issue of regulation of heavy metals in fertilizers" and produce "a study that shows reasonable toxic levels."

Perry, W. E. *Heavy metals & their effects on the recycling of fertilizer products.* The Fertilizer Industry Round Table. TFI, 1200 Pennsylvania Ave., N.W., Washington, D.C. 20469

TFI. *Minutes and agenda items, Heavy Metals Task Force.* (Feb. 20, 1996).
Lohry, Dirk. *Memo to members, TFI Heavy Metals Task Force.* Nutra-Flo Co., Sioux City, IA 51106. (Feb. 23, 1996).

TFI lobbyists in 1994 won removal of a section of a proposed federal Lead Exposure Reduction Act that would have banned fertilizers with more than 0.1 percent lead in the United States. The act's sponsor was never informed about the extent of lead in fertilizer. TFI crowed in an internal memo that "compromise" language removed a general class of consumer products from the limits, thereby exempting fertilizer. The measure passed the Senate and died in the House. (*TFI Legislative and Regulatory Report,* November 1994.)

In 1997, TFI misstepped. Gary Myers, president of TFI, wrote USEPA administrator Carol Browner to complain that the Kansas City EPA office was defining too many sources of metal fume dusts as hazardous wastes. Myers, who is paid $350,000 a year, warned that fertilizer makers would have to raise prices to corn growers unless Browner overturned the Kansas City office. (*Myers letter to Browner [May 9, 1997].*) TFI was soon joined by a larger group of business interests who wrote that the metals and fertilizer industries had always done business under a reading of federal rules claiming that slags, drosses, metal skims, and baghouse dusts were, when recycled, products, not wastes. (*Institute of Scrap Recycling Industries, Steel Manufacturers Association, American Galvanizers Council, Copper Development Association, Copper & Brass Fabricators Council, and Brass & Bronze Ingot Manufacturers letter to Browner [June 10, 1997].*) Thus they would be exempt from hazardous-waste regulation.

While the industry letters were being reviewed at the EPA, "Fear in the Fields" was published July 3–4, 1997. Before long, the industry request was denied by Timothy Fields Jr., acting assistant administrator for the

Office of Solid Waste, writing, "While recycling is clearly one of the goals of the Resource Conservation and Recovery Act, the Agency must ensure that recycling is conducted in a manner that is environmentally sound and protective of human health and the environment." When the industry slags, drosses, skims, and dusts were recycled into fertilizer, Fields wrote, they were used in a manner constituting disposal, thus they were "wastes" (not "products"), thus subject to the hazardous-waste treatment standards. They could still be recycled, but they had to be tested, tracked, and cleaned up to less than 1 part per million cadmium and 5 parts per million lead, the standard test for toxics leaching from hazardous waste. (*Fields letter to TFI et al.* [Aug. 15, 1997] and "Hazardous Waste: Zinc dust industry hit by EPA interpretation of recycling, related to fertilizer controversy," BNA *Daily Environment Report* [Oct. 28, 1997]) The TFI pressure had backfired.

Neither TFI nor the USEPA have ever added up how much hazardous waste went into fertilizer illegally under the industry misinterpretation of the law.

Some members of the TFI board of directors wanted to clean up fertilizer after "Fear in the Fields." They were unaware how much hazardous wastes were used. But publicly, TFI tried to minimize the concern by using absurdly narrow definitions of the terms "hazardous," "waste," "in," and "fertilizer." TFI launched a public-relations campaign marked by half-truths. Other trade groups (chemical retailers, food processors, the Farm Bureau) relied on TFI for information, tantamount to TFI relying on Richard Camp Jr. for information. I was puzzled they weren't more forthcoming; the public trust is precious and fragile. TFI called lead and cadmium "naturally occurring metals," as if they were harmless, occurring in "trace amounts," as if they hadn't been found at 3 percent or higher levels in some products. In 1998, TFI made patently false statements to government officials in a pamphlet distributed to the Association of American Plant Food Control Officials. It said:

FERTILIZER
Does My Fertilizer Contain Hazardous Waste?
Recent news stories have left readers and viewers with the impression fertilizer manufacturers take hazardous waste from industrial generators and put it in fertilizer as a cheap way of disposing of this waste for the generator. Not true. In fact, such a practice would be not only unethical but also illegal.

In fact, hazardous waste recycling into fertilizer was occurring in even the narrowest definitions of those terms, and TFI was hatching plans to minimize the potential regulatory oversight. (TFI staff memos to TFI board of directors, *Fertilizer Risk Assessment Position Paper* [Jan. 26, 1998] and *Background Paper on Proposed Resolution on Heavy Metals* [undated].) Between 1997 and 2000, TFI spent over $1 million for an industry report on heavy metals in fertilizer. (Weinberg Group. *Health risk evaluation of select metals in inorganic fertilizers post application.* TFI, Washington, D.C. [Draft Jan. 11, 2000]) The TFI plan would allow 2,910 parts per million lead in a phosphate fertilizer which, currently, seldom has 30 parts per million lead. Instead of setting a safe floor, TFI was trying to raise the ceiling. Washington State officials said the TFI risk assessment:

- Failed to consider risk from inhaling dust.
- Failed to consider risk from eating beef, dairy, and fish products with heavy metals.
- Failed to consider risk from home garden use of fertilizers, including direct exposure to children.
- Failed to include liming materials despite a USEPA report on "relatively high risk from arsenic in liming agents."
- Failed to answer Rufus Chaney's concerns about long-term cadmium buildup.
- Failed to consider ecological risks such as the effect of metals on soil microbes.
- Failed to include exposure to toxic metals at existing levels, "potentially underestimating risks."

Sorlie, Greg, J. White, and B. Arrington. *Comments to Association of American Plant Food Control Officials (AAPFCO) on California and TFI risk assessments.* Washington Departments of Ecology, Health, and Agriculture, Olympia, WA 98504. [May 15, 2000]).
See also:
Fagin, Dan, Marianne Lavelle, and the Center for Public Integrity. *Toxic Deception: How the chemical industry manipulates science, bends the law, and endangers your health.* Birch Lane Press. 280 pp. (1996).
Shaffer, Matthew. *Waste Lands: The threat of toxic fertilizer.* California Public Interest Research Group. 31 pp. (2001).
Stauber, John, and Sheldon Rampton. *Toxic sludge is good for you! Lies,*

damn lies and the public relations industry. Common Courage Press. 236 pp. (1995).

CHAPTER 9—LAWYERS AND LOSSES

1. Kluck had waged a legal fight with Horsehead Industries, a Pennsylvania hazardous-waste recycler and smelter. Horsehead, like Zinc Nacionale in Mexico, was a full recycler, cleaning the cadmium and lead out of metal waste to produce pure zinc for fertilizer. Unfortunately, that left a toxic mess in and around the factory. See "EPA targets Horsehead recycling," Michael Fabey, *(Allentown, PA) Morning Call* (March 21, 1990); "Company agrees to spend up to $40 million on pollution controls," Associated Press (Aug. 24, 1995); "Settlement by Horsehead Industries Includes $5.65M fines," Dow Jones News Service (Aug. 25, 1995); "Cleaning up in the dark," Melody Peterson, *New York Times* (May 14, 1998). Roger Kluck currently practices law in Seattle.
Perhaps the best practical guide to these laws, aimed at business users, is published by the Bureau of National Affairs. Stimson, James, J. Kimmel, and S. T. Rollin. *Guide to Environmental Laws—From premanufacture to disposal.* BNA Books. 338 pp. (1993).

2. Public Comments. Docket #TMD-00-02-PR, National Organic Program, USDOA, Washington, D.C.
"U.S. to subject organic foods, long ignored, to federal rules." Marian Burros, *New York Times.* (Dec. 15, 1997).

3. *DeYoung v. Cenex and Schaapman,* Case #92-2-00821-8, Superior Court of Washington for Grant County, Ephrata, WA; Case #14802-5-III, Court of Appeals of the State of Washington, Spokane, WA; Case #61959, Supreme Court of the State of Washington, Olympia, WA (1995).
Witte v. Cenex, #CS-97-0413-RHW, U.S. District Court, Yakima, WA 98901.

EPILOGUE

1. Siemens Power Corp. of Richland, Washington, sold 390,000 gallons of ammonium hydroxide containing uranium as fertilizer between 1996 and 2000. The material was a by-product of making nuclear fuel assemblies for commercial power plants worldwide. Germany-based Siemens wanted to empty a waste lagoon. In May 1999, Siemens asked the Nuclear Regulatory Commission to increase the limit on uranium in its recycled waste material from 0.05 parts per million to 1.0 parts per million, a

20-fold increase. Siemens told the NRC it could recycle 540,000 gallons a year as "excellent commercial fertilizer." Washington State officials issued a stop-sale order after learning (from the press, not the NRC) that Siemens was illegally selling the waste to a chemical broker for fertilizer without the testing, disclosure, and licensing required by the new state law. Siemens then applied for a license. Siemens said the average dose of radiation added to a farm family using the product would be 0.000025 millirems a year, less than the radiation from flying in an airline, watching television, or applying phosphate fertilizers. After review, in June 2000, the Washington Department of Agriculture licensed the material.
NRC. *Renewal of Special Nuclear Material License SNM-1227.* Docket 70-1257, NRC, Washington, D.C. 20555.

2. W.R. Grace & Co., of *A Civil Action* infamy, since 1960 has sold a home garden supplement laced with asbestos. Zonolite Chemical Packaging Vermiculite is a puffy material sold in hardware stores as a garden soil supplement. The USEPA found asbestos fibers eighty times higher than worker safety standards. A 1977 memo from Grace executive vice president E. S. Wood shows the company feared a loss of sales if it applied asbestos warning labels to its vermiculite. The memo concluded, "While we have no evidence of adverse effects of our products on consumers, neither can we offer convincing evidence that they are absolutely safe." The USEPA found significant health risks from the natural contaminants in mined vermiculite in 1980, 1985, and 1991, but took no action. Andrew Schneider of the *Seattle Post-Intelligencer* exposed the practice. The USEPA tested three dozen lawn products purchased in eleven states in 2000. The agency found the risk to home gardeners was negligible, but as many as one in one hundred commercial gardeners and nursery workers who used the material regularly could get cancer from it.
Schneider, Andrew. *Seattle Post-Intelligencer.* "Asbestos found in many common garden products." (March 31, 2000) "Asbestos in your garden a hazard?" (Aug. 2, 2000).
Moss, Michael, and A. Appel. *New York Times.* "EPA admits shelving report about asbestos." (July 22, 2000).
Companies in China exported more than a million pounds of highly contaminated zinc used for fertilizer and animal feed in the United States in 1999 and 2000. The shipments of 35 percent zinc sulfate contained a whopping 12 percent cadmium. The contamination was discovered by a Seattle company, triggering a nationwide stop-sale notice on four farm

fertilizers in 2000. The USEPA said the contamination may have come from industrial wastes in the Hunan province of China. The EPA reported, "Preliminary results of those inquiries reveals that as much as 1.3 million pounds entered the U.S. at 10 different ports since the first contaminated shipment was detected in November of 1999. With customs information we have identified 14 importers that have received shipments of suspect zinc sulfate." The importer, Ag-Chem Commission Co., Cornelius, Oregon, quarantined 132 tons at the Port of Seattle and 44 tons at the Port of San Francisco. Idaho authorities tested soil where the material was applied, pronouncing it safely diluted. Two workers who said they were poisoned filed complaints with health authorities. The situation still cries out for further investigation.

Seattle Times, "Tainted fertilizer and feed found," April 6, 2000.
New York Times, "Environmental Protection Agency Seeking Agricultural Products Tainted with Toxic Metal," May 23, 2000.
Memorandum from Michael Stahl, Acting Director, Office of Compliance, EPA, to Deputy Regional Administrators. (Undated).
Recent tests of 19 home and garden fertilizers from around the world found elevated levels of toxic chemicals that may come from industrial waste. Frontier Geosciences accredited laboratory in Seattle found 4.6 ppm arsenic and 1.69 ppm antimony in Vitax fertilizer from England; 98 ppm molybdenum in a product bought at a garden store in Germany; 66 ppm nickel in a Polish fertilizer; 236 ppm chromium, 9 ppm thorium and 1.35 ppm beryllium in Lebanon phosphate fertilizers; and 31,955 ppm aluminum, 377 ppm lead; 269 ppm barium, 19 ppm cadmium, and 2 ppm antimony in a sample from Kerala, India.

3. The principle of precautionary action, endorsed by thirty-one scientists, officials, lawyers, and activists from the United States, Canada, Germany, Britain, and Sweden in 1998, says:

The release and use of toxic substances, the exploitation of resources, and physical alterations of the environment have had substantial unintended consequences affecting human health and the environment. Some of these concerns are high rates of learning deficiencies, asthma, cancer, birth defects and species extinctions, along with global climate change, stratospheric ozone depletion and worldwide contamination with toxic substances and nuclear materials.

We believe existing environmental regulations and other decisions, particu-

larly those based on risk assessment, have failed to protect adequately human health and the environment—the larger system of which humans are but a part.

We believe there is compelling evidence that damage to humans and the worldwide environment is of such magnitude and seriousness that new principles for conducting human activities are necessary.

While we realize that human activities may involve hazards, people must proceed more carefully than has been the case in recent history. Corporations, government entities, organizations, communities, scientists and other individuals must adopt a precautionary approach to all human endeavors.

Therefore, it is necessary to implement the Precautionary Principle: When an activity raises threats of harm to human health or the environment, precautionary measures should be taken even if some cause and effect relationships are not fully established scientifically. In this context the proponent of an activity, rather than the public, should bear the burden of proof.

The process of applying the Precautionary Principle must be open, informed and democratic and must include potentially affected parties. It must also involve an examination of the full range of alternatives, including no action."

4. The Washington Toxics Coalition wanted to limit toxic metals in fertilizer in Washington State to a level no higher than the background level that occurs naturally in clean soil. Industry lobbyists in 1997 said that would be impractical. Industry favored a system like Canada's, allowing toxic metals at application rates that could double the background levels in the soil every forty-five years. The governor and legislature in Washington and other states have so far sided with the industry.

Subsequent analysis showed it might be practical to require fertilizer to be "cleaner than dirt." Of the 2,350 fertilizer analyses submitted to Washington authorities in 1999 and 2000, 727, or almost one-third, passed the test on all nine regulated toxics. Half of the 163 products tested from the Scotts Co., the largest fertilizer distributor to America's hardware and garden stores, passed the test. Most others could pass with minor adjustments.

In general, organic fertilizers, crystallized fertilizers, and nonphosphate fertilizers were the cleanest. Many more phosphate products would pass the test if they were allowed higher cadmium with a ratio of more than 100 to 1 zinc to cadmium, as is recommended by Rufus Chaney, USDA. Personally, I think a requirement that fertilizers be "cleaner than dirt" makes sense, assuring safe recycling and protecting our food-growing soils.

INDEX